Paddy G Farrell

Communications and Coding

RSP is privileged to have Professor Paddy Farrell as Series Editor of the following Series of books:

ELECTRONIC & ELECTRICAL ENGINEERING RESEARCH STUDIES

COMMUNICATIONS SYSTEMS, TECHNIQUES AND APPLICATIONS SERIES

1. Sequence Design for Communications Applications
   Pingzhi Fan *and* **Michael Darnell**

2. Communications Coding and Signal Processing
   THIRD VOLUME ON COMMUNICATION THEORY AND APPLICATIONS
   Edited by **Bahram Honary, Michael Darnell** *and* **Paddy Farrell**

Communications and Coding

Edited by

Michael Darnell
Institute of Integrated Information Systems, University of Leeds, UK

and

Bahram Honary
Communications Research Centre, University of Lancaster, UK

RESEARCH STUDIES PRESS LTD.
Taunton, Somerset, England

JOHN WILEY & SONS INC.
New York · Chichester · Toronto · Brisbane · Singapore

RESEARCH STUDIES PRESS LTD.
24 Belvedere Road, Taunton, Somerset, England TA1 1HD

Marketing and Distribution:

Australia and New Zealand:
JACARANDA WILEY LTD.
Sydney Office, Suite 4A, 113 Wicks Road, North Ryde, NSW 2113, Australia

Canada:
JOHN WILEY & SONS CANADA LIMITED
22 Worcester Road, Rexdale, Ontario, Canada

Europe, Africa, Middle East and Japan:
JOHN WILEY & SONS LIMITED
Baffins Lane, Chichester, West Sussex, UK, PO19 1UD

North and South America:
JOHN WILEY & SONS INC.
605 Third Avenue, New York, NY 10158, USA

South East Asia:
JOHN WILEY & SONS (ASIA) PTE. LIMITED
2 Clementi Loop #02-01
Jin Xing Distripark, Singapore 129809

Library of Congress Cataloging-in-Publication Data

Available

British Library Cataloguing in Publication Data
A catalogue record for this book is available from the British Library.

ISBN 0 86380 238 9 (Research Studies Press Ltd.) *[Identifies the book for orders except in America.]*
ISBN 0 471 98336 5 (John Wiley & Sons Inc.) *[Identifies the book for orders in USA.]*

Printed in Great Britain by SRP Ltd., Exeter

CONTENTS

FOREWORD

This volume contains papers presented at a special Symposium on 'Communications and Coding' held to celebrate the 60th birthday of Paddy Farrell on 25th January 1998. For more than three decades, Paddy has made an outstanding contribution to the fields of digital communications in general, and error control in particular, with many significant publications and invited papers. The Symposium was intended to mark that contribution by bringing together many of his colleagues, collaborators and friends with whom he has interacted over that period to present invited papers on their own current areas of specialisation.

The themes of the Symposium reflected Paddy's specialist interests over the years: error control coding, encryption and sequence design, decoding of error control codes, signal processing in communication systems, and communication systems and networks.

Paddy's interests in error control and digital communication began to develop during his PhD research at the University of Cambridge in the mid-1960s, when he applied simple codes to an atmospheric laser digital data link between the University Library and the Engineering Laboratories - a venture that the somewhat traditional University authorities viewed with a degree of suspicion! From that time, he has been irrevocably 'hooked' on the subject, with a consistent output of high quality contributions - aided by a succession of extremely able research students, several of whom have contributed to the Symposium in their own right. It has truly been a lifetime's commitment.

Many 'generations' of both undergraduate and postgraduate students owe much to Paddy's abilities as a teacher and lecturer. He employs no gimmicks - simply a tremendous enthusiasm for his subject, coupled with a natural ability to be able to express even the most complex concepts in a clear and stimulating manner.

In addition to his specific research achievements, throughout his career Paddy has always given much to the engineering profession in general, both in the UK and wider afield. His long and active involvements with the Institution of Electrical Engineers (IEE), the Institute of Electrical and Electronics Engineers (IEEE), the Institute of Mathematics and its Applications (IMA), the Engineering Council, the Engineering and Physical Sciences Research Council (EPSRC) and NATO are an eloquent testimony of his commitment to the welfare of the engineering profession. His outstanding contribution was recognised in 1991 when he was elected a Fellow of the Royal Academy of Engineering.

Paddy's formative years spent in Argentina, and his dual British/Argentinian nationality, have naturally given an international perspective to his work. In particular, he has always maintained strong collaborative links with South America, the United States of America and mainland Europe. Again, many of these collaborations are reflected in the contributions to the Symposium. He has also had lifelong interests in music and mathematics. In unguarded moments, he has been heard to muse about how his life might have evolved had he chosen mathematics, rather than engineering, as a career path!

Paddy Farrell is a gifted researcher and a delightful friend. It is in a spirit of public acknowledgement of these gifts, and appreciation of his friendship, that this book has been compiled.

Mike Darnell and Bahram Honary
December 1997

CHAPTER 1

Error Control Coding

1. Orthogonal, Antiorthogonal and Self-Orthogonal Matrices and their Codes

James L. Massey
Signal & Information Processing Laboratory
Swiss Federal Institute of Technology
CH-8092 Zürich, Switzerland

Abstract

Orthogonal matrices over arbitrary fields are defined together with their non-square analogs, which are termed row-orthogonal matrices. Antiorthogonal and self-orthogonal square matrices are introduced together with their non-square analogs. The relationships of these matrices to such codes as self-dual codes and linear codes with complementary duals are given. These relationships are used to obtain some constructions of linear codes with complementary duals.

1 Introduction

The aim of this paper is to define a number of different types of matrices over an arbitrary field that are similar to the familiar orthogonal matrices over the real field or to natural extensions of orthogonal matrices, then to show the relationships between these matrices and some familiar linear codes. In particular, these relationships are used to obtain some new constructions of linear codes with complementary duals.

2 Orthogonal Matrices

Let $\mathcal{F}^n$ denote the vector space of n-tuples (or row vectors) with components in an arbitrary field $\mathcal{F}$. The *scalar product* of the vectors $\mathbf{u}$ and $\mathbf{v}$ is the field element $\mathbf{uv}^T$, where here and hereafter the superscripted T denotes transposition. The vectors $\mathbf{u}$ and $\mathbf{v}$ are said to be *orthogonal* when $\mathbf{uv}^T = 0$.

A square matrix $\mathbf{A}$ over $\mathcal{F}$ is said to be *orthogonal* if

$$\mathbf{AA}^T = \mathbf{I},$$

where here and hereafter $\mathbf{I}$ denotes an identity matrix of appropriate dimension. Equivalently, $\mathbf{A}$ is orthogonal just when each row of $\mathbf{A}$ is orthogonal to every *other* row of $\mathbf{A}$ but has a scalar product of 1 with itself.

Note that $\mathbf{A}$ is orthogonal just when $\mathbf{A}^T = \mathbf{A}^{-1}$, and hence $\mathbf{A}^T\mathbf{A} = \mathbf{I}$ so that $\mathbf{A}^T$ is also orthogonal. An orthogonal matrix is not only nonsingular but always has a determinant that is either $+1$ or -1 because $1 = \det(\mathbf{I}) = \det(\mathbf{A}\mathbf{A}^T) = \det(\mathbf{A})\det(\mathbf{A}^T) = (\det(\mathbf{A}))^2$. An example of an orthogonal matrix over the finite field GF(2) is

$$\mathbf{A} = \begin{bmatrix} 0 & 1 & 1 & 1 \\ 1 & 0 & 1 & 1 \\ 1 & 1 & 0 & 1 \\ 1 & 1 & 1 & 0 \end{bmatrix}.$$

Orthogonal matrices over the field $\mathcal{R}$ of real numbers are of great importance in the theory of isometries of $\mathcal{R}^n$, cf. [1].

It seems natural, for an in general non-square, matrix $\mathbf{A}$, to say that $\mathbf{A}$ is *row-orthogonal* if

$$\mathbf{A}\mathbf{A}^T = \mathbf{I},$$

as this is equivalent to the condition that each row of $\mathbf{A}$ is orthogonal to every *other* row of $\mathbf{A}$ but has a scalar product of 1 with itself. A row-orthogonal matrix always has full row rank and thus must have at least as many columns as rows. If $\mathbf{A}$ is row-orthogonal but nonsquare, then $\mathbf{A}^T$ cannot have full row rank and thus cannot also be row-orthogonal. Deleting rows of an orthogonal matrix gives a row-orthogonal matrix, but not every row-orthogonal matrix can be so constructed. For instance, over the field GF(2), the matrix $\mathbf{A} = [1\ 1\ 1]$ is trivially row-orthogonal but there is no orthogonal matrix having $[1\ 1\ 1]$ as a row.

3 Antiorthogonal Matrices

One of the most interesting concepts introduced by P. G. Farrell, in whose honor this paper is written, is that of an *anticode* [2]. A "code" is usually designed to have a large minimum distance between its codewords. Because the opposite of "large minimum distance" is surely "small maximum distance," it was natural for Farrell to use the term *anticode* to describe a set n-tuples designed to have small maximum distance between its "codewords". [We note that sometimes anticodes are defined in such a manner that their "codewords" are all the n-tuples formed by linear combinations of the rows of some matrix, which need not have full row rank so that the "codewords" need not all be different.] The concept of an anticode has found numerous applications both in coding theory and in combinatorics, cf. pp. 548-556 in [3].

Inspired by Farrell's creative terminology, we seek to define an "antiorthogonal matrix" in an appropriate way. Because the opposite of $\mathbf{I}$ is surely $-\mathbf{I}$ [at least if we overlook fields of characteristic 2 for which $\mathbf{I} = -\mathbf{I}$], it seems natural to call a square matrix $\mathbf{B}$ *antiorthogonal* if

$$\mathbf{B}\mathbf{B}^T = -\mathbf{I},$$

i.e., if the rows of $\mathbf{B}$ are pairwise orthogonal but each row has a scalar product of -1 with itself. It follows that $\mathbf{B}$ is antiorthogonal if and only if $\mathbf{B}^{-1} = -\mathbf{B}^T$, and thus $\mathbf{B}^T\mathbf{B} = -\mathbf{I}$ so that $\mathbf{B}^T$ is also antiorthogonal. An example of an antiorthognal matrix over GF(3) is

$$\mathbf{B} = \begin{bmatrix} 1 & 1 \\ 1 & 2 \end{bmatrix}.$$

In a field of characteristic 2, and only in such a field, $-1 = 1$ so that an antiorthogonal matrix is also an orthogonal matrix. Over the real field, the scalar product of a vector with itself is nonnegative, which implies that no antiorthogonal real matrices exist. However, if $\mathbf{A}$ is an orthogonal matrix and i is the imaginary number, then the complex matrix $\mathbf{B} = i\mathbf{A}$ is antiorthogonal.

We now relate antiorthogonal matrices to codes. We first recall that a q-ary code [i.e., a code in which the components of codewords lie in GF(q)] with q^k codewords is *systematic* if it possess an *information set*, i.e., if there is a set of k coordinates such that no two distinct codewords have components that agree in all k of these coordinates. By a permutation of coordinates, which does not affect the Hamming distance between codewords, one obtains an *equivalent code* for which the first k coordinates are an information set, which we shall call a *leading-systematic* code. Every linear code is systematic and hence equivalent to a leading-systematic linear code. Moreover, a linear code is leading-systematic if and only if it has a generator matrix of the form $\mathbf{G} = [\mathbf{I} : \mathbf{P}]$, which generator matrix is easily seen to be unique and is called the *systematic generator matrix* of the code. We recall further that a linear code V is said to be self-dual if $V = V^\perp$ where $V^\perp$ is the dual code of V. If the code length is n, then the dimension of a self-dual code must be $k = n/2$ so that n must be even. We can now give a very simple, but apparently not previously stated, characterization of self-dual codes.

Proposition 1 *A leading-systematic linear code V is self-dual if and only if, in its systematic generator matrix*

$$\mathbf{G} = [\mathbf{I} : \mathbf{P}],$$

the matrix $\mathbf{P}$ is antiorthogonal.

Proof: Because the code length of a self-dual code satisfies $n = 2k$ where k is the code dimension, the matrix $\mathbf{P}$ must be square. Moreover, V will be self-dual just when $\mathbf{G}$ is also a parity-check matrix of the code, i.e., when $\mathbf{G}\mathbf{G}^T = \mathbf{0}$. But $\mathbf{G}\mathbf{G}^T = \mathbf{I} + \mathbf{P}\mathbf{P}^T$ so that V is self-dual just when $\mathbf{P}\mathbf{P}^T = -\mathbf{I}$, as was to be shown.

It seems entirely natural, for an in general nonsquare matrix $\mathbf{B}$, to say that $\mathbf{B}$ is *row-antiorthogonal* if

$$\mathbf{B}\mathbf{B}^T = -\mathbf{I},$$

as this is equivalent to the condition that each row of $\mathbf{B}$ is orthogonal to every *other* row of $\mathbf{B}$ but has a scalar product of -1 with itself. A row-antiorthogonal matrix always has full row rank and thus must have at least as many columns as rows. If $\mathbf{B}$is row-antiorthogonal but nonsquare, $\mathbf{B}^T$ cannot also be row-antiorthogonal. In a field of characteristic 2 and only in such a field, a row-antiorthogonal matrix is also a row-orthogonal matrix. Deleting rows of an antiorthogonal matrix gives a row-antiorthogonal matrix, but not every row-orthogonal matrix can be so constructed, as our previous example of the binary matrix $\mathbf{B} = [1\ 1\ 1]$ demonstrates.

Recalling that a linear code V is said to be *weakly self-dual* if $V \subseteq V^{\perp}$, we obtain a simple generalization of Proposition 1.

Proposition 2 *A leading-systematic linear code V is weakly self-dual if and only if, in its systematic generator matrix*

$$\mathbf{G} = [\mathbf{I}\ :\ \mathbf{P}],$$

the matrix $\mathbf{P}$ is row-antiorthogonal.

Proof: The code V will be self-dual just when the row space of $\mathbf{G}$ is a subset of $V^{\perp}$, i.e., when $\mathbf{G}\mathbf{G}^T = \mathbf{0}$. But $\mathbf{G}\mathbf{G}^T = \mathbf{I} + \mathbf{P}\mathbf{P}^T$ so that V is self-dual just when $\mathbf{P}\mathbf{P}^T = -\mathbf{I}$, i.e., when $\mathbf{P}$ is row-antiorthogonal.

4 Self-Orthogonal Matrices

It seems a natural extension of terminology to say that a square matrix $\mathbf{C}$ over an arbitrary field $\mathcal{F}$ is *self-orthogonal* if

$$\mathbf{C}\mathbf{C}^T = \mathbf{O},$$

where here and hereafter $\mathbf{O}$ denotes a zero matrix of appropriate dimension. Equivalently, $\mathbf{C}$ is self-orthogonal just when each row of $\mathbf{C}$ is orthogonal to *every row* of $\mathbf{C}$ including itself. It follows from $\mathbf{C}\mathbf{C}^T = \mathbf{O}$ that $\det(\mathbf{C}) = 0$ and hence that a self-orthogonal matrix is always singular. An example of a self-orthogonal matrix over the field GF(2) is

$$\mathbf{C} = \begin{bmatrix} 1 & 1 & 1 & 1 \\ 1 & 1 & 1 & 1 \\ 1 & 1 & 0 & 0 \\ 0 & 0 & 1 & 1 \end{bmatrix}.$$

Note that

$$\mathbf{C}^T\mathbf{C} = \begin{bmatrix} 1 & 1 & 0 & 1 \\ 1 & 1 & 0 & 1 \\ 1 & 1 & 0 & 1 \\ 1 & 1 & 0 & 1 \end{bmatrix}$$

so that $\mathbf{C}^T$ is *not* self-orthogonal in this example.

We are now virtually forced to say, for an in general non-square matrix $\mathbf{C}$, that $\mathbf{C}$ is *row-self-orthogonal* if

$$\mathbf{C}\mathbf{C}^T = \mathbf{O},$$

as this is equivalent to the condition that each row of $\mathbf{C}$ is orthogonal to *every row* of $\mathbf{C}$ including itself. A row-self-orthogonal matrix, which is not square (and hence not also a self-orthogonal matrix) can have full row rank. Indeed any matrix obtained by deleting rows from a self-orthogonal matrix is row-self-orthogonal so that deleting the second row from the above-displayed self-orthogonal matrix $\mathbf{C}$ gives a 3×4 matrix that is row-self-orthogonal and has full row rank.

The proofs of Propositions 1 and 2 imply the following alternative characterization of self-dual and weakly self-dual codes.

Proposition 3 *A linear code V with generator matrix $\mathbf{G}$ is self-dual or weakly self-dual if and only if $\mathbf{G}$ is self-orthogonal or row-self-orthogonal, respectively.*

5 Applications to LCD Codes

We now show some connections between the above-defined matrices and *linear codes with complementary duals* (or LCD codes for short). An LCD code is a linear code V such that $V \cap V^{\perp} = \{\mathbf{0}\}$. The reader is referred to [4] for proofs of the basic properties of LCD codes including the fact that if $\mathbf{G}$ is a generator matrix for a linear code V, then V is self-dual if and only if $\mathbf{G}\mathbf{G}^T$ is a nonsingular matrix.

We now show a first connection between LCD codes and the above-defined matrices.

Proposition 4 *A leading-systematic linear code V is an LCD code if (but not only if), in its systematic generator matrix*

$$\mathbf{G} = [\mathbf{I} \,:\, \mathbf{P}],$$

the matrix $\mathbf{P}$ is row-self-orthogonal or, equivalently, if $\mathbf{G}$ is row-orthogonal.

Proof: Because $\mathbf{G}\mathbf{G}^T = \mathbf{I} + \mathbf{P}\mathbf{P}^T$, it follows that $\mathbf{G}$ is row-orthogonal just when $\mathbf{P}$ is row-self-orthogonal. Moreover, if $\mathbf{P}$ is row-self-orthogonal, then $\mathbf{G}\mathbf{G}^T = \mathbf{I}$ so that V is indeed an LCD code.

As an application of Proposition 4, we first note that, for any $k \times m$ matrix $\mathbf{Q}$ over a field of characteristic 2, the $k \times 2m$ matrix $\mathbf{P} = [\mathbf{Q} \,:\, \mathbf{Q}]$ is row-self-orthogonal. Thus $\mathbf{G} = [\mathbf{I} \,:\, \mathbf{Q} \,:\, \mathbf{Q}]$ generates a leading-systematic LCD code of length $n = k + 2m$ and dimension k. In fact, these are the codes used in Proposition 2 of [4] to establish the asymptotic goodness of LCD codes over a finite field of characteristic 2.

More generally, if $\mathbf{Q}$ is any $k \times m$ matrix over a field of characteristic p such that -1 is a quadratic residule modulo p, i.e., such that there exists α in GF(p) for which $\alpha^2 = -1$, then $\mathbf{P} = [\mathbf{Q} \,:\, \alpha\mathbf{Q}]$ is row-self-orthogonal and hence $\mathbf{G} = [\mathbf{I} \,:\, \mathbf{Q} \,:\, \alpha\mathbf{Q}]$ generates a leading-systematic LCD code of length $n = k + 2m$ and dimension k. A theorem of Lagrange (cf. p. 302 in [5]), implies that, for any prime p, one can find elements α, β, γ and δ in GF(p) such that $\alpha^2 + \beta^2 + \gamma^2 + \delta^2 = -1$. The corresponding matrix $\mathbf{P} = [\alpha\mathbf{Q} \,:\, \beta\mathbf{Q} \,:\, \gamma\mathbf{Q} \,:\, \delta\mathbf{Q}]$ is thus row-self-orthogonal. Hence $\mathbf{G} = [\mathbf{I} \,:\, \alpha\mathbf{Q} \,:\, \beta\mathbf{Q} \,:\, \gamma\mathbf{Q} \,:\, \delta\mathbf{Q}]$ generates a leading-systematic LCD code of length $n = k + 4m$ and dimension k. These are the codes used in [4] to establish the asymptotic goodness of LCD codes over an arbitrary finite field.

A stronger consequence of Proposition 4 in the same vein as the previous examples is the following.

Proposition 5 *If* $\mathbf{B}$ *is any* $m \times m$ *antiorthogonal matrix and* $\mathbf{Q}$ *is any* $k \times m$ *matrix, then*

$$\mathbf{G} = [\mathbf{I} \,:\, \mathbf{Q} : \, \mathbf{QB}],$$

is the generator matrix of a leading-systematic LCD code of length $n = k + 2m$ *and dimension* k.

Proof: The proposition follows immediately from Proposition 4 upon noting that $\mathbf{P} = [\mathbf{Q} \,:\, \mathbf{QB}]$ satisfies $\mathbf{PP}^T = \mathbf{QQ}^T + \mathbf{QBB}^T\mathbf{Q}^T = \mathbf{QQ}^T - \mathbf{QQ}^T = \mathbf{O}$ so that $\mathbf{P}$ is indeed a row-self-orthogonal matrix.

The class of codes defined in Proposition 5 is rich enough to meet the asymptotic Varshamov-Gilbert bound as even a crude lower bound on the the number of orthogonal matrices suffices to establish, but we omit details of this argument here.

The following is another consequence of Proposition 4.

Proposition 6 *If* $\mathbf{Q}$ *is any* $k \times k$ *matrix,* $\mathbf{C}$ *is any* $k \times m$ *row-self-orthogonal matrix, and* $\mathbf{A}$ *is any* $m \times m$ *orthogonal matrix, then*

$$\mathbf{G} = [\mathbf{I} \,:\, \mathbf{QCA}],$$

is the generator matrix of a leading-systematic LCD code of length $n = k + m$ *and dimension* k. *The same holds true if* $\mathbf{A}$ *is any* $m \times m$ *antiorthogonal matrix*

Proof: Letting $\mathbf{P} = \mathbf{QCA}$, we have $\mathbf{PP}^T = \mathbf{QCAA}^T\mathbf{C}^T\mathbf{Q}^T = \mathbf{QCC}^T\mathbf{Q}^T = \mathbf{O}$ so that $\mathbf{P}$ is indeed row-self-orthogonal.

6 Closing Remarks

It surpised us somewhat in the course of the work described here that we were able to characterize both self-dual codes and LCD codes so cleanly in terms of row-orthogonal matrices, row-antiorthogonal matrices and row-self-orthogonal matrices. Whether this formulation will prove to be of real use remains to be seen. In [4], it was shown that the complexity of nearest-codeword decoding of an LCD codes is essentially the complexity of the problem, when given a codeword in the dual code $V^{\perp}$, of finding the nearest codeword in the code V. Our suspicion is that the matrix formulations of LCD codes given above may be useful in attacking this problem, and we hope that dexterous coding theorists such as P. G. Farrell will take a crack at confirming our suspicion or showing that it is unfounded..

References

[1] G. A. Jones, "Symmetry," in Handbook of Applicable Mathematics, Vol.5, Combinatorics and Geometry (Eds. W. Lederman and S. Vajda). Chichester and New York: Wiley, 1985, pp. 329-422.

[2] P. G. Farrell, "Linear Binary Anticodes,' *Electronics Letters*, Vol. 6, pp. 419-21, 1970.

[3] F. J. MacWilliams and N. J. A. Sloane, *The Theory of Error Correcting Codes*. Amsterdam: North-Holland, 1977.

[4] J. L. Massey, "Linear Codes with Complementary Duals," *Discrete Math.*, Vol. 106/107, pp. 337-342, 1992. [Also appears as 337-342 in *A Collection of Contributions in Honor of Jack van Lint* (Eds. P. J. Cameron and H. C. A. van Tilborg), Topics in Discrete Math. 7. Amsterdam: Elsevier 1992.]

[5] G. H. Hardy and E. M. Wright, *An Introduction to the Theory of Numbers*. London: Oxford Univ. Press, 4th Ed., 1965.

2. Cascaded Convolutional Codes

Stefan Höst[(1)], Rolf Johannesson[(1)], Vladimir R. Sidorenko[(2)], Kamil Sh. Zigangirov[(1)], and Viktor V. Zyablov[(2)]

[(1)] Dept. of Information Technology
Lund University
P.O. Box 118
S-221 00 Lund, Sweden
stefanh@it.lth.se
rolf@it.lth.se
kamil@it.lth.se

[(2)] Inst. for Problems of Information Transmission
of the Russian Academy of Science
B. Karetnyi per., 19, GSP-4
Moscow, 101447 Russia
sid@ippi.ac.msk.su
zyablov@ippi.ac.msk.su

Abstract—Concatenation is a both powerful and practical method to obtain constructions that are attractive for use in communication situations where very low error probabilities are needed. In this paper the simplest such construction is considered, viz., a cascade of two convolutional encoders without an interleaver. For simplicity we consider only the case when the rates of the outer and inner encoders are matched. The active distances are lower-bounded and a counterpart to Costello's lower bound on the free distance is derived for the ensemble of periodically time-varying cascaded convolutional codes. Some structural properties are shown, e.g., that the cascade of two minimal encoders is minimal. Finally, the active distances are lower-bounded for time-invariant cascaded convolutional codes.

I. Introduction

Recently there has been a dramatic increase in the interest in concatenated schemes with conventional convolutional codes as constituent codes. Most concatenated schemes include an interleaver.

In this paper we consider the simplest concatenated scheme with two convolutional encoders without an interleaver but with matched rates. We will show that there exist

*This research was supported in part by the Swedish Research Council for Engineering Sciences under Grant 94–83 and in part by the Royal Swedish Academy of Sciences in liaison with the Russian Academy of Sciences.

cascaded convolutional codes with good free distance properties. From the construction follows that they might be attractive to use in various communication situations.

The cascaded convolutional codes are introduced in Section II. In Section III we give a brief review of the active distances. Section IV is devoted to ensemble properties. The definitions of the active distances given in Section III are generalized to time-varying convolutional codes and lower-bounded for the ensemble of time-varying cascaded convolutional codes. It is shown that Costello's lower bound on the free distance also holds for our cascaded scheme. In Section V we study the structural properties of the cascaded convolutional codes and in Section VI we lower-bound various distances for time-invariant cascaded convolutional codes.

II. Cascaded Convolutional Codes—A First Encounter

Consider a binary, rate $R = b/c, b \leq c$, convolutional code with a $b \times c$ polynomial generator matrix $G(D)$ of memory m. The causal information (input) sequence

$$\boldsymbol{u}(D) = \boldsymbol{u}_0 + \boldsymbol{u}_1 D + \boldsymbol{u}_2 D^2 + \dots, \tag{1}$$

where $\boldsymbol{u}_i = u_i^{(1)} \dots u_i^{(b)}$, $i = 0, 1, \dots$, is encoded as the causal codeword (output)

$$\boldsymbol{v}(D) = \boldsymbol{v}_0 + \boldsymbol{v}_1 D + \boldsymbol{v}_2 D^2 + \dots, \tag{2}$$

where $\boldsymbol{v}_i = v_i^{(1)} \dots v_i^{(c)}$, $i = 0, 1, \dots$, and where

$$\boldsymbol{v}(D) = \boldsymbol{u}(D) G(D). \tag{3}$$

For simplicity we sometimes write $\boldsymbol{u} = \boldsymbol{u}_0 \boldsymbol{u}_1 \dots$ and $\boldsymbol{v} = \boldsymbol{v}_0 \boldsymbol{v}_1 \dots$ instead of $\boldsymbol{u}(D)$ and $\boldsymbol{v}(D)$, respectively. Then, we replace (3) by

$$\boldsymbol{v} = \boldsymbol{u}\boldsymbol{G}, \tag{4}$$

where the semi-infinite matrix $\boldsymbol{G}$ is given by

$$\boldsymbol{G} = \begin{pmatrix} G_0 & G_1 & \dots & G_m & & & \\ & G_0 & G_1 & \dots & G_m & & \\ & & \ddots & \ddots & & & \ddots \end{pmatrix}, \tag{5}$$

where G_i, $0 \leq i \leq m$, is a binary $b \times c$ matrix. The cascade of two convolutional encoders without an interleaver but with matched rates is a cascade of a rate $R_o = b_o/c_o$ outer polynomial encoder of memory m_o and a rate $R_i = b_i/c_i$ inner polynomial encoder of memory m_i, where $b_i = c_o$. The *cascaded convolutional code* $\mathcal{C}_c$ is encoded by the rate $R_c = R_o R_i = b_o/c_i \stackrel{\text{def}}{=} b_c/c_c$ convolutional encoder of memory $m_c \leq m_o + m_i$ as shown in Fig. 1.

Let the information sequence $\boldsymbol{u}^c$ of the cascaded encoder $\boldsymbol{G}^c$ be encoded as the cascaded codeword $\boldsymbol{v}^c$, where

$$\boldsymbol{v}^c = \boldsymbol{u}^c \boldsymbol{G}^c. \tag{6}$$

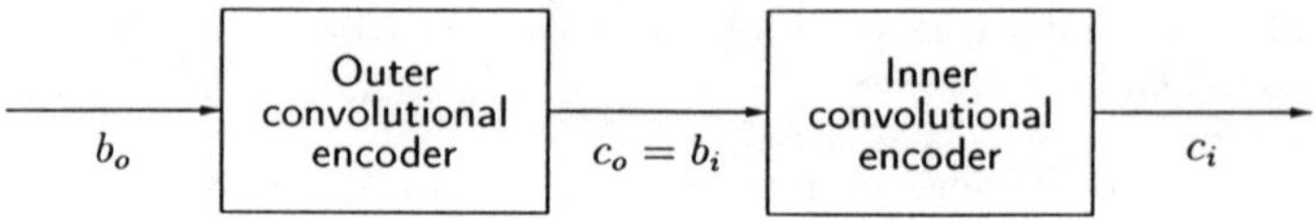

Figure 1: A cascade of two consecutive convolutional encoders.

The cascaded generator matrix $\boldsymbol{G}^c$ is given by

$$\boldsymbol{G}^c = \boldsymbol{G}^o \boldsymbol{G}^i, \tag{7}$$

where $\boldsymbol{G}^o$ and $\boldsymbol{G}^i$ are the generator matrices of the outer and inner encoders, respectively. Since we assume that $b_i = c_o$, it follows that (7) could, equivalently, be written

$$G^c(D) = G^o(D)G^i(D). \tag{8}$$

When the rates of the constituent convolutional codes do not match, i.e., when $b_i \neq c_o$, then the multiplication in (8) is not defined, but, somewhat surprisingly, (7) is still valid.

III. Active Distances—A Primer

The active distances were introduced in [1]; see also [2], [3]. Here we give a brief summary. Consider a binary, rate $R = b/c$ convolutional code of memory m. Let the binary m-dimensional vector of b-tuples $\boldsymbol{\sigma}_t$ be the (physical) state at depth t of a realization in controller canonical form of the generator matrix and let $\boldsymbol{\sigma}_t^{(i)}$ be the b-tuple representing the contents of position i of the shift registers (counted from the input connections). It follows that $\boldsymbol{\sigma}_t = \boldsymbol{\sigma}_t^{(1)}\boldsymbol{\sigma}_t^{(2)} \ldots \boldsymbol{\sigma}_t^{(m)}$. To the information sequence $\boldsymbol{u} = \boldsymbol{u}_0\boldsymbol{u}_1 \ldots$ corresponds a state sequence $\boldsymbol{\sigma} = \boldsymbol{\sigma}_0\boldsymbol{\sigma}_1 \ldots$.

Let $\mathcal{S}_{[t_1,t_2]}^{\sigma_1,\sigma_2}$ denote the set of state sequences $\boldsymbol{\sigma}_{t_1}, \boldsymbol{\sigma}_{t_1+1}, \ldots, \boldsymbol{\sigma}_{t_2}$ that start in state $\boldsymbol{\sigma}_1$ and terminate in state $\boldsymbol{\sigma}_2$ and do not have two consecutive allzero states in between, i.e.,

$$\mathcal{S}_{[t_1,t_2]}^{\sigma_1,\sigma_2} \stackrel{\text{def}}{=} \{\boldsymbol{\sigma}_{t_1}, \boldsymbol{\sigma}_{t_1+1}, \ldots, \boldsymbol{\sigma}_{t_2} \mid \boldsymbol{\sigma}_{t_1} = \boldsymbol{\sigma}_1, \boldsymbol{\sigma}_{t_2} = \boldsymbol{\sigma}_2 \text{ and } \boldsymbol{\sigma}_i, \boldsymbol{\sigma}_{i+1} \text{ not both} = \boldsymbol{0},\ t_1 \leq i < t_2\}. \tag{9}$$

Definition: Let $\mathcal{C}$ be a convolutional code encoded by a polynomial generator matrix $G(D)$ of memory m. The *jth order active row distance* is

$$a_j^r = \min_{\boldsymbol{u}_j \neq \boldsymbol{0},\, \mathcal{S}_{[0,j+1]}^{\boldsymbol{0},\sigma}} \left\{ w_H\left(\boldsymbol{u}_{[0,j]}\boldsymbol{G}_j^r\right)\right\}, \tag{10}$$

where $\boldsymbol{\sigma}$ denotes any value of the state $\boldsymbol{\sigma}_{j+1}$ and

$$\boldsymbol{G}_j^r = \begin{pmatrix} G_0 & G_1 & \ldots & G_m & & & \\ & G_0 & G_1 & \ldots & G_m & & \\ & & \ddots & \ddots & & \ddots & \\ & & & G_0 & G_1 & \ldots & G_m \end{pmatrix} \tag{11}$$

is a $(j+1) \times (j+1+m)$ truncated version of the semi-infinite matrix $\boldsymbol{G}$ given in (5). □

Definition: Let $\mathcal{C}$ be a convolutional code encoded by a polynomial generator matrix $G(D)$ of memory m. The *jth order active column distance* is

$$a_j^c = \min_{\mathcal{S}_{[0,j+1]}^{0,\sigma}} \left\{ w_H \left(\boldsymbol{u}_{[0,j]} \boldsymbol{G}_j^c \right) \right\}, \tag{12}$$

where $\boldsymbol{\sigma}$ denotes any encoder state and

$$\boldsymbol{G}_j^c = \begin{pmatrix} G_0 & G_1 & \dots & G_m & & & & \\ & G_0 & G_1 & \dots & G_m & & & \\ & & \ddots & \ddots & & \ddots & & \\ & & & G_0 & G_1 & \dots & G_m \\ & & & & G_0 & & G_{m-1} \\ & & & & & \ddots & \vdots \\ & & & & & & G_0 \end{pmatrix} \tag{13}$$

is a $(j+1) \times (j+1)$ truncated version of the semi-infinite matrix $\boldsymbol{G}$. □

Definition: Let $\mathcal{C}$ be a convolutional code encoded by a polynomial generator matrix $G(D)$ of memory m. The *jth order active reverse column distance* is

$$a_j^{rc} = \min_{\mathcal{S}_{[m,m+j+1]}^{\sigma,0}} \left\{ w_H \left(\boldsymbol{u}_{[0,j+m]} \boldsymbol{G}_j^{rc} \right) \right\}, \tag{14}$$

where $\boldsymbol{\sigma}$ denotes any encoder state and

$$\boldsymbol{G}_j^{rc} = \begin{pmatrix} G_m & & & \\ G_{m-1} & G_m & & \\ \vdots & G_{m-1} & \ddots & \\ G_0 & \vdots & & G_m \\ & G_0 & & G_{m-1} \\ & & \ddots & \vdots \\ & & & G_0 \end{pmatrix} \tag{15}$$

is a $(j+m+1) \times (j+1)$ truncated version of the semi-infinite matrix $\boldsymbol{G}$. □

Definition: Let $\mathcal{C}$ be a convolutional code encoded by a polynomial generator matrix $G(D)$ of memory m. The *jth order active segment distance* is

$$a_j^s = \min_{\mathcal{S}_{[m,m+j+1]}^{\sigma_1,\sigma_2}} \left\{ w_H \left(\boldsymbol{u}_{[0,j+m]} \boldsymbol{G}_j^s \right) \right\}, \tag{16}$$

where $\boldsymbol{\sigma}_1$ and $\boldsymbol{\sigma}_2$ denote any encoder states, and $\boldsymbol{G}_j^s = \boldsymbol{G}_j^{rc}$. □

The jth order active row distance is characterized by a fixed number of "free" information tuples, $j+1$, followed by a varying number, between $\nu_{\min}$ and m, of "zero state" driving information tuples. Sometimes we find it useful to consider a corresponding distance between two paths of fixed total length, $j+1$, but with a varying number of "free" information tuples. Hence we introduce our final active distance:

Definition: Let $\mathcal{C}$ be a convolutional code encoded by a polynomial generator matrix $G(D)$ of memory m. The *jth order active burst distance* is

$$a_j^b = \min_{\mathcal{S}_{[0,j+1]}^{0,0}} \left\{ w_H(\boldsymbol{u}_{[0,j]} \boldsymbol{G}_j^c) \right\}, \tag{17}$$

where $\boldsymbol{G}_j^c$ is given in (13). □

The active row and burst distances are related via the following inequalities:

$$\begin{cases} a_j^b & \geq \min_i \{a_{j-\nu_i}^r\} \\ a_j^r & \geq \min_i \{a_{j+\nu_i}^b\}. \end{cases} \tag{18}$$

Clearly, when $\nu_{\text{min}} = m$, we have

$$a_j^b = \begin{cases} \text{undefined}, & 0 \leq j < m \\ a_{j-m}^r, & j \geq m. \end{cases} \tag{19}$$

For a noncatastrophic generator matrix we have

$$\min_j \{a_j^b\} = d_{\text{free}}. \tag{20}$$

The active distances can be lower-bounded by the following affine functions:

$$\begin{cases} a_j^r & \geq \alpha j + \beta^r \\ a_j^c & \geq \alpha j + \beta^c \\ a_j^{rc} & \geq \alpha j + \beta^{rc} \\ a_j^s & \geq \alpha j + \beta^s \\ a_j^b & \geq \alpha j + \beta^b. \end{cases} \tag{21}$$

Let $\mathcal{A}$ denote the set of the values of the parameters $\{\alpha, \beta^r, \beta^c, \beta^{rc}, \beta^s, \beta^b\}$ that optimize the lower bounds (21).

Example 1:

In Fig. 2 the active distances for the encoding matrix

$$G(D) = \begin{pmatrix} 1 + D + D^2 + D^3 + D^5 & 1 + D^2 + D^3 + D^5 \end{pmatrix} \tag{22}$$

are shown together with the functions defined by $\mathcal{A}$, where

$$\mathcal{A} = \left\{ \alpha = \frac{1}{3}, \beta^r = \frac{19}{3}, \beta^c = 2, \beta^{rc} = 1, \beta^s = -1, \beta^b = \frac{14}{3} \right\}. \tag{23}$$

The difference between β^{rc} and β^c is due to the fact that the distance profile of the reverse encoding matrix, $D^5G(D^{-1})$, is not as good as the distance profile of $G(D)$. □

IV. Time-varying Cascaded Convolutional Codes

So far we have considered only *time-invariant* or *fixed* convolutional codes, i.e., convolutional codes encoded by time-invariant generator matrices. When it is too difficult to

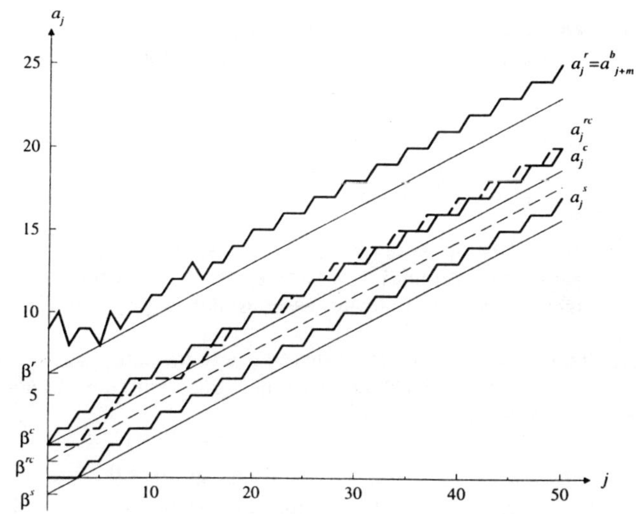

Figure 2: The active distances for the encoding matrix in Example 1.

analyze the performance of a communication system using time-invariant convolutional codes, we can often obtain powerful results if we study time-varying convolutional codes instead.

Assuming polynomial generator matrices we have

$$\boldsymbol{v}_t = \boldsymbol{u}_t G_0 + \boldsymbol{u}_{t-1} G_1 + \ldots + \boldsymbol{u}_{t-m} G_m, \tag{24}$$

where G_i, $0 \leq i \leq m$, is a binary $b \times c$ time-invariant matrix.

In general a binary convolutional code can be *time-varying*. Then (24) becomes

$$\boldsymbol{v}_t = \boldsymbol{u}_t G_0(t) + \boldsymbol{u}_{t-1} G_1(t) + \ldots + \boldsymbol{u}_{t-m} G_m(t), \tag{25}$$

where $G_i(t), i = 0, 1, \ldots, m$, is a binary $b \times c$ time-varying matrix.

Remark: With a slight abuse of terminology we call for simplicity a time-varying polynomial transfer funtion matrix a *generator matrix* although it might not have full rank.

We have the general *ensemble of binary, rate $R = b/c$, time-varying convolutional codes* with generator matrices of memory m in which each digit in each of the matrices $G_i(t)$ for $0 \leq i \leq m$ and $t = 0, 1, 2, \ldots$ is chosen independently and is equally likely to be 0 and 1.

As a special case of the ensemble of time-varying convolutional codes we have the ensemble of binary, rate $R = b/c$, *periodically* time-varying convolutional codes encoded by a time-varying polynomial generator matrix of memory m and *period* T,

$$\boldsymbol{G}_t = \begin{pmatrix} G_0(t) & G_1(t+1) & \ldots & G_m(t+m) & & \\ & G_0(t+1) & G_1(t+2) & \ldots & G_m(t+1+m) & \\ & & \ddots & \ddots & & \ddots \end{pmatrix}, \tag{26}$$

in which each digit in each of the matrices $G_i(t) = G_i(t+T)$ for $0 \leq i \leq m$ and $t = 0, 1, \ldots, T-1$ is chosen independently and is equally likely to be 0 and 1. We denote this ensemble $\mathcal{EV}(b, c, m, T)$.

The ensemble $\mathcal{EI}(b, c, m)$ of *time-invariant* convolutional codes can be considered as the special case $\mathcal{EV}(b, c, m, 1)$ and the ensemble of general time-varying convolutional codes defined above will be denoted $\mathcal{EV}(b, c, m, \infty)$.

Before we define the active distances for periodically, time-varying convolutional codes encoded by time-varying polynomial generator matrices we introduce the following sets of information sequences, where we always assume that $t_1 \leq t_2$.

Let $\mathcal{U}^r_{[t_1-m, t_2+m]}$ denote the set of information sequences $\boldsymbol{u}_{t_1-m}\boldsymbol{u}_{t_1-m+1}\ldots\boldsymbol{u}_{t_2+m}$ such that the first m and the last m subblocks are zero and such that they do not contain $m+1$ consecutive zero subblocks, i.e.,

$$\begin{aligned}\mathcal{U}^r_{[t_1-m,t_2+m]} \stackrel{\text{def}}{=} \{&\boldsymbol{u}_{[t_1-m,t_2+m]} \mid \boldsymbol{u}_{[t_1-m,t_1-1]} = \boldsymbol{0},\\ &\boldsymbol{u}_{[t_2+1,t_2+m]} = \boldsymbol{0}, \text{ and } \boldsymbol{u}_{[i,i+m]} \neq \boldsymbol{0}, t_1 - m \leq i \leq t_2\}.\end{aligned} \tag{27}$$

Let $\mathcal{U}^c_{[t_1-m,t_2]}$ denote the set of information sequences $\boldsymbol{u}_{t_1-m}\boldsymbol{u}_{t_1-m+1}\ldots\boldsymbol{u}_{t_2}$ such that the first m subblocks are zero and such that they do not contain $m+1$ consecutive zero subblocks, i.e.,

$$\begin{aligned}\mathcal{U}^c_{[t_1-m,t_2]} \stackrel{\text{def}}{=} \{&\boldsymbol{u}_{[t_1-m,t_2]} \mid \boldsymbol{u}_{[t_1-m,t_1-1]} = \boldsymbol{0},\\ &\text{and } \boldsymbol{u}_{[i-m,i]} \neq \boldsymbol{0}, t_1 \leq i \leq t_2\}.\end{aligned} \tag{28}$$

Let $\mathcal{U}^{rc}_{[t_1-m,t_2]}$ denote the set of information sequences $\boldsymbol{u}_{t_1-m}\boldsymbol{u}_{t_1-m+1}\ldots\boldsymbol{u}_{t_2}$ such that the last m subblocks are zero and such that they do not contain $m+1$ consecutive zero subblocks, i.e.,

$$\begin{aligned}\mathcal{U}^{rc}_{[t_1-m,t_2]} \stackrel{\text{def}}{=} \{&\boldsymbol{u}_{[t_1-m,t_2]} \mid \boldsymbol{u}_{[t_2-m+1,t_2]} = \boldsymbol{0},\\ &\text{and } \boldsymbol{u}_{[i-m,i]} \neq \boldsymbol{0}, t_1 \leq i \leq t_2\}.\end{aligned} \tag{29}$$

Let $\mathcal{U}^s_{[t_1-m,t_2]}$ denote the set of information sequences $\boldsymbol{u}_{t_1-m}\boldsymbol{u}_{t_1-m+1}\ldots\boldsymbol{u}_{t_2}$ such that they do not contain $m+1$ consecutive zero subblocks, i.e.,

$$\mathcal{U}^s_{[t_1-m,t_2]} \stackrel{\text{def}}{=} \{\boldsymbol{u}_{[t_1-m,t_2]} \mid \boldsymbol{u}_{[i-m,i]} \neq \boldsymbol{0}, t_1 \leq i \leq t_2\}. \tag{30}$$

Next we introduce the $(j+m+1) \times (j+1)$ truncated, time-varying generator matrix:

$$\boldsymbol{G}_{[t,t+j]} = \begin{pmatrix} G_m(t) & & & \\ G_{m-1}(t) & G_m(t+1) & & \\ \vdots & G_{m-1}(t+1) & \ddots & \\ G_0(t) & \vdots & \ddots & G_m(t+j) \\ & G_0(t+1) & & G_{m-1}(t+j) \\ & & \ddots & \vdots \\ & & & G_0(t+j) \end{pmatrix}. \tag{31}$$

We are now well-prepared to generalize the definitions of the active distances for convolutional codes encoded by polynomial generator matrices to time-varying convolutional codes encoded by time-varying polynomial generator matrices. When we consider ensembles of time-varying convolutional codes we assume that $\nu_{\min} = m$, and, hence, the active burst distance is simply a shift of the active row distance. Thus, in the sequel we do not consider the active burst distance separately.

Definition: Let $\mathcal{C}$ be a time-varying convolutional code encoded by a time-varying polynomial generator matrix of memory m.

The *jth order active row distance* is

$$a_j^r \stackrel{\text{def}}{=} \min_t \min_{\mathcal{U}^r_{[t-m,t+j+m]}} \{w_H(\boldsymbol{u}_{[t-m,t+j+m]}\boldsymbol{G}_{[t,t+j+m]})\}. \tag{32}$$

The *jth order active column distance* is

$$a_j^c \stackrel{\text{def}}{=} \min_t \min_{\mathcal{U}^c_{[t-m,t+j]}} \{w_H(\boldsymbol{u}_{[t-m,t+j]}\boldsymbol{G}_{[t,t+j]})\}. \tag{33}$$

The *jth order active reverse column distance* is

$$a_j^{rc} \stackrel{\text{def}}{=} \min_t \min_{\mathcal{U}^{rc}_{[t-m,t+j]}} \{w_H(\boldsymbol{u}_{[t-m,t+j]}\boldsymbol{G}_{[t,t+j]})\}. \tag{34}$$

The *jth order active segment distance* is

$$a_j^s \stackrel{\text{def}}{=} \min_t \min_{\mathcal{U}^s_{[t-m,t+j]}} \{w_H(\boldsymbol{u}_{[t-m,t+j]}\boldsymbol{G}_{[t,t+j]})\}. \tag{35}$$

□

For a time-varying convolutional code encoded by a time-varying, noncatastrophic, polynomial generator matrix with active row distance a_j^r we define its free distance by

$$d_{\text{free}} \stackrel{\text{def}}{=} \min_j\{a_j^r\}. \tag{36}$$

Consider the ensemble $\mathcal{EVC}(b_c, c_c, m_c, T_c)$ of periodically time-varying, cascaded convolutional codes constructed in the following way:

Choose as an outer convolutional code a binary, time-invariant, rate $R_o = b_o/c_o$ convolutional code encoded by a minimal-basic encoding matrix of memory $m_o = \min_k\{\nu_{o,k}\}$, where $\nu_{o,k}$ is the constraint length of the left input, and whose starting point of its active segment distance is

$$j_s^o = \frac{R_o}{1-R_o} m_o + O(\log_2 m_o). \tag{37}$$

The existence of such a convolutional code was shown in [1].

The ensemble of inner convolutional codes is the ensemble of binary, periodically time-varying with period T_i, rate $R_i = b_i/c_i$ convolutional codes encoded by time-varying polynomial generator matrices of memory $m_i \geq j_s^o$.

Clearly, we have

$$T_c = T_i. \tag{38}$$

Theorem 1 *Consider a periodically time-varying, rate $R_c = b_c/c_c$, cascaded convolutional code encoded by a convolutional encoder of memory m_c.*

(i) *Let the information sequences be restricted to the set $\mathcal{U}^r_{[t-m_c,t+j+m_c]}$. Then the code symbols in the segment $\boldsymbol{v}_{[t,t+j+m_c]}$ are mutually independent and equiprobable over the ensemble $\mathcal{EVC}(b_c, c_c, m_c, T_c)$ for all j, $0 \le j < T_c$.*

(ii) *Let the information sequences be restricted to the set $\mathcal{U}^c_{[t-m_c,t+j]}$. Then the code symbols in the segment $\boldsymbol{v}_{[t,t+j]}$ are mutually independent and equiprobable over the ensemble $\mathcal{EVC}(b_c, c_c, m_c, T_c)$ for all j, $0 \le j < T_c$.*

(iii) *Let the information sequences be restricted to the set $\mathcal{U}^{rc}_{[t-m_c,t+j]}$. Then the code symbols in the segment $\boldsymbol{v}_{[t,t+j]}$ are mutually independent and equiprobable over the ensemble $\mathcal{EVC}(b_c, c_c, m_c, T_c)$ for all j, $0 \le j < T_c$.*

(iv) *Let the information sequences be restricted to the set $\mathcal{U}^s_{[t-m_c,t+j]}$. Then the code symbols in the segment $\boldsymbol{v}_{[t,t+j]}$ are mutually independent and equiprobable over the ensemble $\mathcal{EVC}(b_c, c_c, m_c, T_c)$ for all j, $0 \le j < T_c$.* □

Proof: Analogously to the proof of the corresponding theorem for the ensemble $\mathcal{EV}(b, c, m, T)$ of periodically time-varying convolutional codes given in [1]. □

If we let

$$l = (j+1)/m_c, \tag{39}$$

then we obtain asymtotically—for large memories m_c—the following counterpart to the corresponding theorem for the ensemble of periodically time-varying convolutional codes $\mathcal{EV}(b, c, m, T)$ [1]:

Theorem 2

(i) *There exists a cascaded convolutional code in the ensemble $\mathcal{EVC}(b_c, c_c, m_c, T_c)$ whose normalized active row distance asymptotically satisfies*

$$\delta^r_l \overset{\text{def}}{=} \frac{a^r_j}{m_c c} \ge h^{-1}\left(1 - \frac{l}{l+1}R_c\right)(l+1) + O\left(\frac{\log_2 m_c}{m_c}\right) \tag{40}$$

for $l \ge l_0 = O\left(\frac{1}{m_c}\right)$.

(ii) *There exists a cascaded convolutional code in the ensemble $\mathcal{EVC}(b_c, c_c, m_c, T_c)$ whose normalized active column distance asymptotically satisfies*

$$\delta^c_l \overset{\text{def}}{=} \frac{a^c_j}{m_c c} \ge h^{-1}(1 - R_c)l + O\left(\frac{\log_2 m_c}{m_c}\right) \tag{41}$$

for $l \ge l_0 = O\left(\frac{\log_2 m_c}{m_c}\right)$.

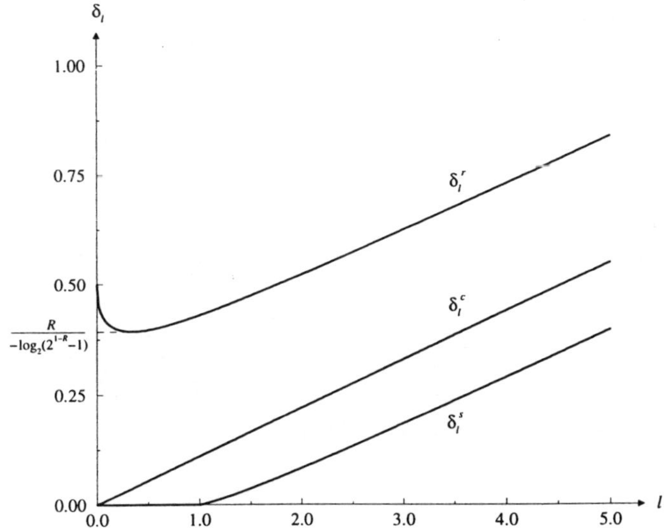

Figure 3: Typical behavior of the lower bounds on the normalized active distances of Theorem 2.

(*iii*) *There exists a cascaded convolutional code in the ensemble $\mathcal{EVC}(b_c, c_c, m_c, T_c)$ whose normalized active segment distance asymptotically satisfies*

$$\delta_l^s \stackrel{\text{def}}{=} \frac{a_j^s}{m_c c_c} \geq h^{-1}(1 - \frac{l+1}{l} R_c) l + O\left(\frac{\log_2 m_c}{m_c}\right) \tag{42}$$

for $l \geq l_s^o = \frac{R_o}{1-R_o} + O\left(\frac{\log_2 m_o}{m_o}\right)$.

□

The typical behavior of the bounds in Theorem 2 is shown in Fig. 3. Notice that by minimizing the lower bound on the normalized active row distance we obtain nothing but the main (normalized) term in Costello's lower bound on the free distance, viz.,

$$\frac{R}{-\log_2(2^{1-R} - 1)}.$$

The free distance for a convolutional code is obtained as the minimum weight of the nonzero codewords. The only restriction on the input sequence is that it should be nonzero. Now, let us consider the cascade in Fig. 1. Clearly, also the free distance for the cascaded convolutional code, d_{free}^c, is obtained as the minimum weight of the nonzero codewords; again the minimum is evaluated over all nonzero inputs. Since the inputs to the inner encoder are restricted to be codewords of the outer encoder we will not obtain a useful estimate of d_{free}^c from the free distance of the inner code, d_{free}^i.

It is somewhat surprising that, given only a restriction on the memory of the inner code, there exists a convolutional code obtained as a simple cascade with a free distance satisfying the Costello bound [4]:

Theorem 3 (Costello bound) *Consider a cascade $\mathcal{C}_c$ of an outer binary, periodically time-varying convolutional code $\mathcal{C}_o$ of rate $R_o = b_o/c_o$ and encoder memory m_o and an inner binary, time-varying convolutional code $\mathcal{C}_i$ of rate $R_i = b_i/c_i$ and encoder memory m_i, where $b_i = c_o$. If*

$$m_i \geq \frac{R_o}{1-R_o} m_o, \tag{43}$$

then there exists a pair of codes, $\mathcal{C}_o$ and $\mathcal{C}_i$, such that the code $\mathcal{C}_c$ of rate $R_c = R_o R_i$ and encoder memory $m_c = m_o + m_i$ has a free distance satisfying the inequalities

$$\begin{aligned} d_{free}^c &\geq \frac{m_c c_i R_c}{-\log_2(2^{1-R}-1)} + O(\log_2 m_c) \\ &\geq \frac{m_o c_o R_o}{-\log_2(2^{1-R_o}-1)} + \frac{m_i c_i R_i}{-\log_2(2^{1-R_i}-1)} \\ &+ O(\log_2 m_o) + O(\log_2 m_i) \end{aligned} \tag{44}$$

□

Proof: Choose an inner code $\mathcal{C}_i$ such that

$$m_i = \frac{R_o}{1-R_o} m_o. \tag{45}$$

Then, by minimizing (40), we obtain

$$\begin{aligned} d_{\text{free}}^c &\geq \frac{m_c b_o}{-\log_2(2^{1-R_c}-1)} + O(\log_2 m_c) \\ &= \frac{m_c c_i R_c}{-\log_2(2^{1-R_c}-1)} + O(\log_2 m_c) \\ &= \frac{m_o b_o}{-\log_2(2^{1-R_c}-1)} + \frac{m_i c_i R_c}{-\log_2(2^{1-R_c}-1)} + O(\log_2 m_c) \\ &\geq \frac{m_o c_o R_o}{-\log_2(2^{1-R_o}-1)} + \frac{m_i c_i R_i}{-\log_2(2^{1-R_i}-1)} + O(\log_2 m_c), \end{aligned} \tag{46}$$

where we in the last inequality have used the fact that

$$-\log_2(2^{1-R}-1)$$

is increasing and

$$\frac{R}{-\log_2(2^{1-R}-1)}$$

is decreasing functions of R. □

Theorem 3 shows that, given the restriction on the ratio m_i/m_o, from the Costello lower bound point of view we lose nothing in free distance by splitting a given amount of convolutional encoder memory into two cascaded convolutional encoders!

Remark: Assume that for the cascaded encoder in Theorem 3 we have $\nu_{\min,o} = m_o$ and $\nu_{\min,i} = m_i$ and that $m_i = \frac{R_o}{1-R_o} m_o$ holds. Then the total number of states in the outer and inner encoders are

$$\begin{aligned} 2^{m_o b_o} + 2^{m_i b_i} &= 2^{m_o b_o}(1 + 2^{m_i b_i - m_o b_o}) \\ &= 2^{(m_o + m_i) b_o}(1 + 2^{-m_i b_o}), \end{aligned} \tag{47}$$

which is essentially equal to the total number of states of a generator matrix $G^c(D)$ with $\nu_{\min,c} = m_c$. The second equality follows from the equality $m_i = \frac{R_o}{1-R_o} m_o$.

We have shown that the performances for all active distances for the ensemble of time-varying cascaded convolutional codes is the same as for the ensemble of time-varying convolutional codes although the former ensemble is essentially smaller.

V. Structural Properties

In this section we will show some structural properties for time-invariant cascaded convolutional generator matrices.

Let $G^c_{mb}(D)$ be a minimal-basic [5] [6] encoding matrix equivalent to the cascaded generator matrix

$$G^c(D) = G^o(D)G^i(D) \tag{48}$$

and denote by m_{mb} and ν_{mb} its memory and the constraint length, respectively. An obvious realization of the cascaded generator matrix $G^c(D)$ is to realize both $G^o(D)$ and $G^i(D)$ on controller canonical form. Such a realization requires $\nu_o + \nu_i$ memory elements. Since a realization of a minimal-basic encoding matrix on controller canonical form will require a minimum number of memory elements among all realizations of equivalent encoders for the code, the constraint length of $G^c_{mb}(D)$ must be less than or equal to the number of memory elements in this obvious realization. Thus we have

Theorem 4 *Let $G^c_{mb}(D)$ be a minimal-basic encoding matrix equivalent to the cascaded generator matrix $G^c(D)$ defined by the product of the two basic matrices $G^o(D)$ and $G^i(D)$, then*

$$\nu_{mb} \leq \nu_o + \nu_i. \tag{49}$$

□

To state a similar theorem for the memory of a cascaded generator matrix we need

Lemma 5 *If $G^o(D)$ and $G^i(D)$ are two basic encoding matrices, then the corresponding cascaded encoding matrix, $G^c(D)$, is also basic.* □

Proof: Since $G^o(D)$ and $G^i(D)$ are basic, they are both polynomial and have polynomial right inverses. Their product will, of course, be polynomial, and the cascade will have a polynomial right inverse, viz.,

$$(G^c(D))^{-1} = (G^o(D)G^i(D))^{-1} = (G^i(D))^{-1}(G^o(D))^{-1}. \tag{50}$$

□

We will now prove the following

Theorem 6 *Let $G^c_{mb}(D)$ be a minimal-basic encoding matrix equivalent to the cascaded generator matrix, $G^c(D)$, defined by the product of the two basic matrices $G^o(D)$ and $G^i(D)$, then*

$$m_{mb} \leq m_o + m_i. \tag{51}$$

□

Proof: Since the constraint lengths are invariants of equivalent minimal-basic encoding matrices [6], it follows that $m_{mb} \leq m_c$, where m_c is the memory of a cascaded generator matrix and m_{mb} is the memory of an equivalent minimal-basic encoding matrix. Furthermore, the product of two polynomial matrices cannot have entries of larger degrees than the sum of the maximum degrees of each of the two matrices. Hence, we conclude that

$$m_{mb} \leq m_c \leq m_o + m_i. \tag{52}$$

□

Example 2:

Consider the rate $R_o = 2/3$ outer encoding matrix [7]

$$G^o(D) = \begin{pmatrix} 1 & D & 1+D \\ D^2 & 1 & 1+D+D^2 \end{pmatrix}, \tag{53}$$

of memory $m_o = 2$ and constraint length $\nu_o = 3$, and the rate $R_i = 3/4$ inner encoding matrix [7]

$$G^i(D) = \begin{pmatrix} 1 & 1 & 1 & 1 \\ 0 & 1+D & D & 1 \\ 0 & D & 1+D^2 & 1+D^2 \end{pmatrix}, \tag{54}$$

of memory $m_i = 2$ and constraint length $\nu_i = 3$.

The corresponding cascaded encoding matrix is

$$\begin{aligned} G^c(D) &= G^o(D)G^i(D) \\ &= \begin{pmatrix} 1 & 1 & D+D^3 & D^2+D^3 \\ D^2 & 1+D^3 & 1+D^2+D^3+D^4 & D+D^2+D^3+D^4 \end{pmatrix}. \end{aligned} \tag{55}$$

The cascaded encoding matrix has rate $R_c = 2/4$, memory $m_c = 4 = m_o + m_i$, and constraint length $\nu_c = 7 > \nu_o + \nu_i$.

Both $G^o(D)$ and $G^i(D)$ are minimal-basic, while $G^c(D)$ is only basic. As expected, the product of two minimal-basic encoding matrices is not necessarily minimal-basic.

We have thc following equivalent minimal-basic encoding matrix

$$G^c_{mb}(D) = \begin{pmatrix} 1 & 1 & D+D^3 & D^2+D^3 \\ D+D^2 & 1+D+D^3 & 1+D^3 & D+D^2 \end{pmatrix} \tag{56}$$

with $m_{mb} = 3$ and $\nu_{mb} = 6$. Comparing these parameters with the memories and constraint lengths for the outer and inner encoding matrices we get $m_{mb} = 3 < m_o + m_i = 4$ and $\nu_{mb} = \nu_o + \nu_i = 6$, and, thus, the upper bound on the constraint lengths in Theorem 4 is tight. □

Since the constraint length for $G^c(D)$ is $\nu_c = 7 \not\leq \nu_o + \nu_i = 6$ it follows that the requirement that $G^c_{mb}(D)$ is minimal-basic in Theorem 4 is essential.

The encoding matrix $G^c(D)$ in (55) is not minimal-basic but the next theorem shows that it is in fact minimal.

Theorem 7 *If $G^o(D)$ and $G^i(D)$ are two minimal encoding matrices, then the cascaded encoding matrix defined by their product is also minimal.* □

Proof: The *span* of a set of $f(D)$ of Laurent series in D is the interval from the index of the first nonzero component of $f(D)$ to the index of the last nonzero component, if there is one, or to infinity otherwise. In [8] it is proved that a rational generator matrix $G(D)$ is minimal if and only if the span of the information sequence is contained in the span of the code sequence, i.e., if and only if,

$$\text{span}(\boldsymbol{u}(D)) \subseteq \text{span}(\boldsymbol{u}(D)G(D)). \tag{57}$$

Thus, if both the outer and inner generator matrices are minimal, then we have

$$\text{span}(\boldsymbol{u}^c(D)) \subseteq \text{span}(\boldsymbol{u}^c(D)G^o(D)) \subseteq \text{span}(\boldsymbol{u}^c(D)G^o(D)G^i(D)) = \text{span}(\boldsymbol{u}^c(D)G^c(D)), \tag{58}$$

where the first inclusion is equivalent to the minimality of $G^o(D)$ and the second inclusion is equivalent to the minimality of $G^i(D)$. The minimality of $G^c(D)$ is equivalent to (58) and the proof is complete. □

Example 3:

Consider the rate $R_o = 1/2$ outer encoding matrix

$$G^o(D) = \begin{pmatrix} 1+D & D \end{pmatrix} \tag{59}$$

of memory $m_o = 1$ and constraint length $\nu_0 = 1$ and the rate $R_i = 2/3$ inner encoding matrix

$$G^i(D) = \begin{pmatrix} 1 & 1 & D \\ 1+D & D & 1 \end{pmatrix} \tag{60}$$

of memory $m_i = 1$ and constraint length $\nu_i = 2$.

Both encoding matrices are minimal-basic and, in contrast to Example 2, so is their product

$$G^c_{mb}(D) = G^c(D) = \begin{pmatrix} 1+D^2 & 1+D+D^2 & D^2 \end{pmatrix}. \tag{61}$$

Here $\nu_{mb} = 2 < \nu_o + \nu_o = 3$, but $m_{mb} = m_o + m_i = 2$, and for these encoding matrices the upper bound on the memories in Theorem 6 is tight. □

VI. Lower Bounds on Active Distances for Time-invariant Cascaded Convolutional Codes

The active distances can be lower-bounded by the following affine functions:

$$\begin{cases} a_j^r & \geq & \alpha j + \beta^r & = & f^r(j) \\ a_j^c & \geq & \alpha j + \beta^c & = & f^c(j) \\ a_j^{rc} & \geq & \alpha j + \beta^{rc} & = & f^{rc}(j) \\ a_j^s & \geq & \alpha j + \beta^s & = & f^s(j), \end{cases} \tag{62}$$

where α is the asymptotic slope of the active distances and the β:s are chosen as large as possible. It is clear that $\beta^r \geq \beta^c \geq \beta^s$ and $\beta^r \geq \beta^{rc} \geq \beta^s$. The active distances are in general difficult to calculate. In this section we will derive lower bounds on the active distances for time-invariant cascaded convolutional codes.

We assume that both the outer and inner polynomial encoding matrices are minimal. Then it follows from Theorem 7 that also the cascaded encoding matrix is minimal. The cascaded encoder is considered to be in the allzero state if both the inner and outer encoders are in the allzero state.

Define a *burst of zeros* as a sequence of zero symbols from the outer encoder that drives the inner encoder to the zero state. The length of such sequence to the inner encoder is lower-bounded by

$$l_{burst} \geq \min_j\{\nu_i(j)\}\frac{c_o}{b_i}, \tag{63}$$

where $\nu_i(j)$ is the constraint length for the jth input of the inner encoder. Whenever the outer encoder must not get to the zero state the length of the burst can also be upper-bounded by

$$l_{burst} \leq \left\lfloor \frac{j_s^o c_o + 2(c_o - 1)}{b_i} \right\rfloor, \tag{64}$$

where j_s^o is the longest sequence of zero symbols from the outer encoder. Call this maximum length l_b. From the following three lemmas we get some knowledge of what is happening on the output of the outer encoder. They lower-bound the output weight before and after a burst of zeros of maximum length.

Lemma 8 *The number of ones, w_b, between the beginning of the output sequence and the first burst of zeros, occupying l_b symbols, must satisfy*

$$w_b \geq a^c_{l_b+\lceil \frac{w_b}{c} \rceil -1}. \tag{65}$$

□

Proof: The weight of the output sequence, from the beginning of the sequence to the end of the zero burst, is w_b. Therefore the subsequence before the l_b zero symbols has weight w_b. This subsequence must consist of at least $\lceil w_b/c \rceil$ symbols. Assume that k is the true length of the sequence, then

$$k \geq l_b + \left\lceil \frac{w_b}{c} \right\rceil. \tag{66}$$

Since the active column distance is a nondecreasing function it is clear that

$$a^c_{l_b+\lceil \frac{w_b}{c} \rceil -1} \leq a^c_{k-1} \leq w_b \tag{67}$$

□

Lemma 9 *The number of ones, w_e, between the last burst of zeros, occupying l_b symbols, and the end of the output sequence must satisfy:*

$$w_e \geq a^{rc}_{l_b+\lceil \frac{w_e}{c} \rceil -1}. \tag{68}$$

□

Proof: Similar to the proof of Lemma 8 the weight of the output sequence from the beginning of the last burst to the end of the sequence is w_e and the weight of the sequence following the burst is w_e. The length of this subsequence is at least $\lceil w_{e,j}/c \rceil$. Again let k be the true length of the information sequence. Since the active reverse column distance is a nondecreasing function we have

$$a^{rc}_{l_b+\lceil \frac{w_e}{c} \rceil -1} \leq a^{rc}_{k-1} \leq w_e. \tag{69}$$

□

Lemma 10 *The number of ones, w_s, between two bursts of zeros, occupying l_b symbols each, must satisfy:*

$$w_s \geq a^s_{2l_b+\lceil \frac{w_s}{c} \rceil -1}. \tag{70}$$

□

Proof: The weight of the sequence is w_s, and the length must be at least $2l_b + \lceil w_s/c \rceil$. Let k be the true length. Since the active segment distance is nondecreasing we have

$$a^s_{2l_b+\lceil \frac{w_s}{c} \rceil -1} \leq a^s_{k-1} \leq w_s. \tag{71}$$

□

To simplify the notations let j_i denote the number of information symbols within the span of the input to the inner encoder when the number of information symbols within the span of the input to the outer encoder is j. When the outer encoder has the predictable degree property this number is lower-bounded by

$$j_i \geq \left\lceil (j + \nu^o_{\min}) \frac{c_o}{b_i} \right\rceil, \tag{72}$$

where $\nu^o_{\min}$ is the minimum constraint length of the outer encoder. When the cascaded encoder is not in the zero state we might still have either the outer or the inner encoder in the zero state, but not both at the same time. Assuming that the inner encoder is not in the zero state, then the minimum output weight from the inner encoder will be $a^{r,i}_{j_i}$ both when the outer encoder is allowed to reach the zero state and when it is not. Thus, there are two different cases to consider:

i) The inner encoder is not in the zero state.

ii) The inner encoder is in the zero state, and the outer is not.

From (63) and (64) we know the range for the length of a zero sequence that will drive the inner encoder to the zero state. Since the output from the encoder is zero, while in zero state, the burst should be chosen as long as possible, i.e., l_b. Define φ_b as the reduction in Hamming weight made when a burst of zeros is inserted. If the burst starts after k information symbols this reduction will be (Fig. 4)

$$\begin{aligned} \varphi_b &= f^r_i(j_i) - (f^r_i(k-1) + f^r_i(j_i - k - l_b)) = \\ &= \alpha(l_b + 1) + \beta^r. \end{aligned} \tag{73}$$

If φ_b is positive, then we have the same reduction for a second burst of zeros. Finally, assume that the sequence has been split by N_r bursts of zeros. The length of the subsequences is for the

- first subsequence lower-bounded by $l \geq \lceil w_b/b_i \rceil$
- last subsequence lower-bounded by $l \geq \lceil w_e/b_i \rceil$
- middle subsequences lower-bounded by $l \geq \lceil w_s/b_i \rceil$

and, consequently, the largest possible number of bursts is the largest N_r such that

$$N_r l_b + (N_r - 1) \left\lceil \frac{w_s}{b_i} \right\rceil + \left\lceil \frac{w_b}{b_i} \right\rceil + \left\lceil \frac{w_e}{b_i} \right\rceil \leq j_i + 1, \tag{74}$$

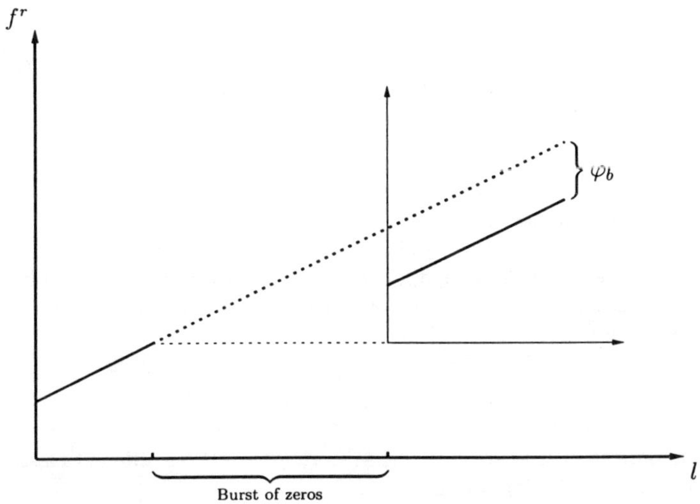

Figure 4: One burst of zeros splitting the sequence into two subsequences.

The active row distance for the cascaded encoder can be lower-bounded by

$$a_j^{r,c} \geq \min\left\{f_i^r(j_i),\ f_i^r\left(\left\lceil \frac{w_b}{b_i} \right\rceil - 1\right) + (N_r - 1) f_i^r\left(\left\lceil \frac{w_s}{b_i} \right\rceil - 1\right) + f_i^r\left(\left\lceil \frac{w_e}{b_i} \right\rceil - 1\right)\right\} \tag{75}$$

and its free distance by

$$d_\infty^c \geq \min_j \{a_j^{r,c}\}. \tag{76}$$

When calculating the active column distance and the active reverse column distance we start by considering the case when there are no bursts of zeros and the inner encoder is not driven to the zero state. Then the sequence is split by one burst of zeros. If the output weight is reduced, the same reduction will be made for every new inserted burst. To get the maximum number of bursts when calculating $a_j^{c,c}$ we shall find the largest N_c such that

$$N_c l_b + N_c \left\lceil \frac{w_s}{b_i} \right\rceil + \left\lceil \frac{w_b}{b_i} \right\rceil \leq j_i + 1. \tag{77}$$

Similarly, when lower-bounding $a_j^{rc,c}$, the maximum number of bursts is given by the largest N_{rc} satisfying

$$N_{rc} l_b + N_{rc} \left\lceil \frac{w_s}{b_i} \right\rceil + \left\lceil \frac{w_e}{b_i} \right\rceil \leq j_i + 1. \tag{78}$$

The bounds on the active column distance and active reverse column distance for the cascaded code can be calculated as

$$a_j^{c,c} \geq \min\left\{f_i^c(j_i),\ f_i^r\left(\left\lceil \frac{w_b}{b_i} \right\rceil - 1\right) + (N_c - 1) f_i^r\left(\left\lceil \frac{w_s}{b_i} \right\rceil - 1\right) + f_i^c\left(\left\lceil \frac{w_s}{b_i} \right\rceil - 1\right)\right\} \tag{79}$$

and

$$a_j^{rc,c} \geq \min\left\{f_i^{rc}(j_i),\ f_i^{rc}\left(\left\lceil\frac{w_s}{b_i}\right\rceil - 1\right) + (N_{rc}-1)f_i^{r}\left(\left\lceil\frac{w_s}{b_i}\right\rceil - 1\right) + f_i^{r}\left(\left\lceil\frac{w_e}{b_i}\right\rceil - 1\right)\right\}, \quad (80)$$

respectively.

The first case to consider for the active segment distance is, as before, the case when there are no bursts. For the second case one burst is introduced and the sequence is split into two subsequences where the weight of the first is the corresponding active column distance, and the weight of the second is the corresponding active reverse column distance. Each of these subsequences should be split further. The maximum number of bursts is the largest N_s such that

$$N_s l_b + N_s \left\lceil\frac{w_s}{b_i}\right\rceil \leq j_i + 1. \quad (81)$$

Thus, to calculate the active segment distance there are three different cases to consider, which results in the following formula

$$\begin{aligned} a_j^{s,c} \geq \min\Bigg\{ & f_i^{c}(j_i),\ f_i^{rc}\left(\left\lceil\frac{w_b}{b_i}\right\rceil - 1\right) + f_i^{c}\left(j_i - \left\lceil\frac{w_e}{b_i}\right\rceil - l_b\right), \\ & f_i^{rc}\left(\left\lceil\frac{w_b}{b_i}\right\rceil - 1\right) + (N_s-1)f_i^{r}\left(\left\lceil\frac{w_s}{b_i}\right\rceil - 1\right) + f_i^{c}\left(\left\lceil\frac{w_e}{b_i}\right\rceil - 1\right)\Bigg\}. \end{aligned} \quad (82)$$

VII. Comments

In this paper we have studied cascaded convolutional codes with matched rates, i.e., the rates of the constituent convolutional codes are restricted by $b_i = c_o$. In a forthcoming paper we will consider cascades with unmatched rates, i.e., we will allow $b_i \neq c_o$ [9].

References

[1] S. Höst, R. Johannessson, K. Sh. Zigangirov, and V. V. Zyablov, "Active Distances for Convolutional Codes", *submitted to IEEE Trans. Inform. Theory.*

[2] J. Justesen, C. Thommesen, and V. V. Zyablov, "Concatenated codes with convolutional inner codes", *IEEE Trans. Inform. Theory,* vol. IT-34, pp. 1217–1225, Sept. 1988.

[3] C. Thommesen and J. Justesen, "Bounds on distances and error exponents of unit-memory codes", *IEEE Trans. Inform. Theory,* vol. IT-29, pp. 637–649, July 1983.

[4] D. Costello, "Free Distance Bounds for Convolutional Codes", *IEEE Trans. Inform. Theory,* IT-20:356–366, 1974.

[5] G. D. Forney, Jr., "Convolutional codes I: Algebraic structure", *IEEE Trans. Inform. Theory,* IT-16:720–738, Nov. 1970.

[6] R. Johannessson and Z. Wan, "A Linear Algebra Approach to Minimal Convolutional Encoders", *IEEE Trans. Inform. Theory*, IT-39:1219–1233, July 1993.

[7] L. Perez and D. Costello, "Cascaded Convolutional Codes", *proc. 1995 IEEE Whistler, Canada*, pp. 160, Sept. 1995.

[8] G. D. Forney, Jr., R. Johannesson, and Z. Wan, "Minimal and Canonical Rational Generator Matrices for Convolutional Codes", *IEEE Trans. Inform. Theory*, IT-42:1865–1880, Nov. 1996.

[9] S. Höst, R. Johannesson, V. R. Sidorenko, K. Sh. Zigangirov, and V. V. Zyablov, "A Note on Cascaded Convolutional Codes with Unmatched Rates", in preparation.

3. Randomness and Structure in Channel Coding

Gérard Battail
Ecole Nationale Supérieure des Télécommunications
Département Communications and URA 820 of CNRS,
46, rue Barrault,
F-75634 Paris Cedex 13, France

Abstract

Codes are sought so that their distance distribution mimics that of random coding. Depending on how the closeness of the distance distribution with respect to that of random coding is specified, we distinguish strongly and weakly random-like codes. The latter may have good bit (or symbol) error rate but their word error rate is poor. A crude model of them is proposed. Attempting to design random-like convolutional codes leads to recursive codes where the recursive part of the encoder is specified by a primitive polynomial. Although their decoding is difficult due to the large memory needed for good performance, their use as components in turbo codes is a simple and efficient means for obtaining weakly random-like codes with a specified bit error rate at an information rate close to the channel capacity. Moreover, the many-component turbo code scheme asymptotically appears as intrinsically random-like regardless of the component codes.

1 Introduction

Shannon introduced for proving the channel coding theorem the extraordinary idea of "random coding" [1]. Roughly stated, if you choose a code at random, then you most probably get a good code. Thus, the problem of channel coding appears at first sight as trivial: any code is good with high probability.

However, this optimistic view does not resist a naive attempt for implementing random coding. The problem lies at the receiver side where, in the absence of any structure, one has to compare *each* received word with *all* possible codewords. Since moreover the channel coding theorem is asymptotic in the code length n, the codebook contains $M = 2^{nR}$ codewords for keeping a constant information rate R. Hence its size grows exponentially with the codeword length n, which need be large for good performance.

Should we conclude that random coding is practically impossible, where these words are given the very strong meaning that an exponentially growing decoding complexity in terms of the code length is unavoidable? The answer is "yes" if we attempt to actually construct a code at random, but "no" if instead of a code *generated by random means*, we look for a code which *mimics* random coding. The only reason why a code generated by random coding is good lies in its distance properties. Any deterministic code with a distance spectrum close to that obtained in the average by random coding is a good code. In other words, we have to look for codes which share some properties with random coding, although they are generated by deterministic rules and not by chance. These codes will be endowed with some structure so their decoding will be tractable.

As a comparison, let us think of *pseudo-random* sequences which are known for many years. They are generated by simple, deterministic means, although they mimic the average frequency and correlation properties of actual random sequences. Their usefulness in generating noise, for instance, is recognized for long and these sequences are widely and successfully used in the simulation of communication systems as well as of many other systems where chance has a role. Similarly, we may aim at finding deterministic codes which mimic random coding.

2 Looking for random-like codes

We thus proposed a few years ago [2] to search for *random-like* codes instead of using the minimum distance criterion which dominated the development of "error correcting codes", especially when algebraic construction methods were used. Several families of "random-like codes" were identified, including well known and practically efficient codes like the Reed-Solomon ones [3], but we believe that the most important example of random-like codes is that of *turbo codes* [4]. Although the reason for their outstanding performance remains controversial and further work is still needed before it is fully understood, we believe that the features of turbo codes are best explained if we interpret them as random-like codes. Moreover, besides a better understanding this approach may provide rules for the explicit design of turbo codes with still improved performance.

Paradoxically, pseudo-random coding has even an advantage with respect to true random coding. Drawing a code at random only succeeds with high probability (if the code is long enough) in a good distance spectrum. However, a random-like construction always succeeds in it, since it endows the sequences it generates with the average properties of a random code. According to the MacMillan theorem, sequences generated by random means can be separated into typical sequences and atypical ones, where the probability that a sequence belongs to the second group decreases when the sequence length increases and asymptotically approaches zero. On the contrary a random-like construction, since it results in sequences endowed with the average properties of random coding, always generates a typical-like sequence.

To begin with, we shall make more explicit the idea of random-like coding by looking at the criterion which specifies the closeness of the distance distribution of a code with respect to that obtained in the average by random coding. We shall be led to distinguish two broad classes of random-like codes, to be referred to as strongly and weakly random-like [5]. The *strongly random-like codes* exhibit a good minimum distance, so they achieve a small word error rate. In the perspective of an infinitely increasing word length, however, the word error rate appears as a performance measure which gets tighter and tighter as the length increases. From an engineering point of view, the bit or symbol error rate is a more natural way of specifying the performance of a code, and it is almost always used in the technique. It turns out that the codes of the other type namely, the *weakly random-like codes*, can achieve some specified bit or symbol error rate so they are better fitted to the engineering needs although their word error rate is rather poor. The extensive use which has been made of the minimum distance criterion makes this domain almost unexplored. We believe that most of the outstanding features of turbo codes may be explained by the fact they actually belong to the family of weakly random-like codes.

We shall show that weakly random-like convolutional codes, to be referred to as *pseudo-random*, may be designed [6, 7]. Their decoding remains however a difficult problem, but we shall show that their interleaved concatenation i.e., their use as components in turbo codes, results in excellent performance [7, 8]. Moreover, we shall show that even if the component codes of a turbo code are not exactly of this type, the turbo code scheme itself may be interpreted as a means to make the distance spectrum approach that of random coding as the number of component codes increases. Any specified performance at a rate less than the channel capacity can thus in principle be achieved by simply increasing this number.

3 Strongly vs weakly random-like codes

3.1 Strongly random-like codes

How may we express the closeness of the distance spectrum of a code with respect to that obtained in the average by random coding? We shall for the sake of simplicity restrict ourselves to binary linear codes. Then, the distance spectrum reduces to the weight spectrum. Let us assume that a linear block code of length n has

$$A(z) = 1 + A_1 z + \cdots + A_w z^w + \cdots + A_n z^n$$

as weight enumerator polynomial. This means that the code, besides the word made of n 0's, contains A_w words of weight $w > 0$. We have obviously $\sum_{i=0}^{n} A_i = 2^k$, with $A_0 = 1$.

We may first think of demanding that *each* of the integers A_w, $w > 0$, is a close approximation of the average number R_w of words of weight w obtained by random coding with the same parameters (n, k). This case is referred to as *strongly random-like*. We have actually to take account of the fact that some of the A_w may be zero as a consequence of the code structure. For instance, in many cases $A_w = 0$ for w odd. Then, we shall assume that A_w for w even is an approximation of e.g. $R_{w-1} + R_w$. In the case of a binary (n, k) code, we have

$$R_w = 2^{-(n-k)} \binom{n}{w},$$

since a single word has weight w with probability $2^{-n}\binom{n}{w}$ and the code contains 2^k words. It is in general a rational number, but for defining a strongly random-like code, we look for an integer approximation of it, for instance to the nearest integer. Of course, we shall have $A_w = 0$ for $R_w < 1/2$. The numbers A_w will thus be 0 for the smallest values of w, so the code has a minimum weight d_s (the subscript s stands here for "strongly") i.e., the smallest value of w such that $A_w > 0$.

It is an easy matter to show that a strongly random-like code meets the Gilbert-Varshamov bound, regardless of the particular criterion which is used for specifying the approximation (see Appendix A). The Gilbert-Varshamov bound is a lower bound on the largest possible minimum distance of a code, so a strongly random-like code is good for the conventional minimum distance criterion, too. Thus, if it were possible to construct strongly random-like codes, they would be good in the conventional sense. Unfortunately, they are even harder to construct than good codes for the minimum distance criterion, since they imply a very good minimum distance in addition to other properties, especially that of having very few words of small weights. We show in Appendix B that the capacity of the additive Gaussian white noise channel cannot be reached if one just looks for packing spheres in the n-dimensional Euclidean space. This is intended to refute the minimum distance criterion as a means for approaching capacity, which geometrically would lead to look for such a packing. The use of random coding (or the attempt to mimic its results which we designated by random-like coding) is actually a means for very evenly distributing points in a large-dimensional space, whereas no deterministic means is known to this end.

3.2 Weakly random-like codes

The closeness of a weight distribution with respect to that obtained in the average by random coding can however be expressed by other means than in Section 3.1. First of all, let us consider the normalized weight distributions of the given code and of random coding, where normalizing means that the number of codewords of a certain weight w is divided by the total number of codewords, namely 2^k. These are

$$a_w = A_w 2^{-k}$$

and

$$r_w = R_w 2^{-k} = 2^{-n}\binom{n}{w}.$$

We obviously have

$$\sum_{w=0}^{n} a_w = \sum_{w=0}^{n} r_w = 1,$$

so we may deal with such normalized weight coefficients as if they were probabilities. The sets of such normalized coefficients may be dealt with as probability distributions and they will be denoted by $a \triangleq \{a_w\}$ and $r \triangleq \{r_w\}$. Then, a very natural closeness measure between such distributions is the *cross-entropy*, also known as Kullback divergence, namely:

$$H(a\|r) \triangleq \sum_{w=0}^{n} a_w \log(a_w/r_w), \tag{1}$$

a positive quantity if $a \neq r$ as a consequence of the Gibbs inequality. Therefore, a weakly random-like code may be defined by the condition $H(a\|r) < \varepsilon$, where ε is some given positive constant. Similarly to the case of strongly random-like codes, we must actually take care of the possible cancellation of certain weight coefficients due to the code structure e.g., of the odd-weight ones. In this particular case, r_w should be replaced in (1) by $r'_w = r_{w-1} + r_w$ or by $r'_w = 2r_w$. Denoting by r' the modified distribution associated with random coding i.e., normalized in terms of the non-zero coefficients of a, the weakly random-like codes are thus defined by their normalized weight distribution which obeys:

$$H(a\|r') < \varepsilon. \tag{2}$$

Interestingly, this inequality only implies a shape similarity the closer, the smaller is ε. Since a_w factors the logarithm in (1), the closeness may be much stronger in the central part of the distribution, where r_w is large, than in its tails where r_w is small. Thus, (2) may obtain even if the minimum weight d of the code is much smaller than the minimum weight d_s of the corresponding strongly random-like code, provided that a_w is small enough for $d \leq w \leq d_s$. In order to understand the properties of a weakly random-like code, we shall first introduce a crude model of it and examine its bit error rate. Then, a more realistic extension of it will be discussed. We shall later see that the properties of such models are similar to those of turbo codes, which will substantiate our claim that they are actually weakly random-like codes.

3.3 A crude model of weakly random-like codes, and its analysis

As a crude model of weakly random-like codes [5], let us consider a linear (n, k) code with a weight distribution equal to the true normalized distribution of random coding, multiplied by a coefficient close to 1, say $1 - \eta$, plus a single coefficient $\eta \ll 1$ at a weight $d \ll d_s$. This code has thus an almost random-like weight distribution, except for a very small fraction $\eta = 2^{-k} A_d$ of all codewords which has the small weight d. Then, using the union bound on the bit error rate (BER) and the approximations stated in Appendix C, we may represent the BER performance of the model as shown in Fig. 1 where the natural logarithm of an upper bound on the bit error rate, denoted by y, is plotted as a function of the channel signal-to-noise ratio γ (with a linear scale). As shown in Appendix C, it can be approximated by the broken line which comprises, from left to right:

a horizontal floor at $y = \ln(1/2)$, for $0 \leq \gamma \leq \gamma_C$, where γ_C is the signal-to-noise ratio for which the channel capacity is achieved, since the channel is almost useless when the capacity is exceeded;

a very steeply descending curve, which corresponds to the performance of random coding. It has been represented by a vertical line at $\gamma = \gamma_C$ i.e., n is assumed to be very large;

a horizontal line at $y = \ln(d/n)$ since d/n is an obvious upper bound of the BER;

and an oblique line of slope $-d$, according to (15), which cuts the y axis at $k \ln 2 + \ln \eta = \ln A_d$, a non-negative quantity.

If d is small enough and n large enough, then a small constant bit error rate d/n occurs in a range of signal-to-noise ratios between γ_C and some value beyond which the bit error rate slowly decreases with slope $-d$ in terms of γ. Fig. 2 contains two examples

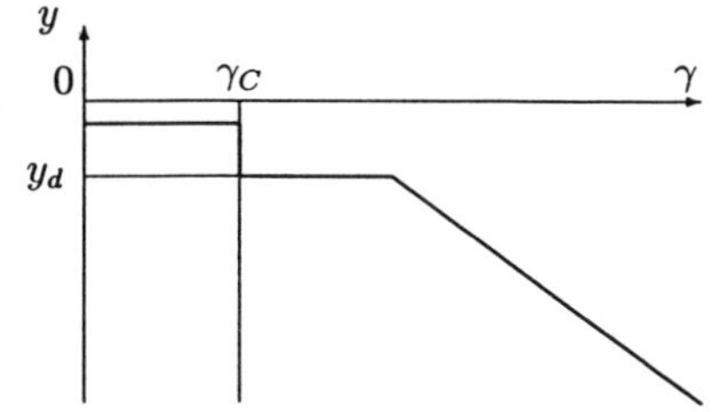

Figure 1: Broken line representation of the union upper bound on the bit error rate for the model of Sect. 3.3. The natural logarithm y of a single term of the bound is plotted in ordinate while the signal-to-noise ratio γ of the channel is plotted in abscissa with a linear scale. The actual bound curve is close to the thick broken line, where the leftmost horizontal line corresponds to $y = \ln(1/2)$ for $\gamma < \gamma_C$, the leftmost very steep line (actually drawn as vertical) represents the performance of random coding, the rightmost line has slope $-d$, and the horizontal line in the middle at $y_d = \ln(d/n)$ corresponds to the limiting BER d/n.

drawn assuming the following numerical values: both codes have length $n = 1,024$, rate $R = 1/2$, a single word of weight d i.e., $A_d = 1$, but different minimum distances $d = 1$ and $d = 2$.

As an extension of this crude model, we may think of a linear weakly random-like code as generated by a matrix of the form

$$\mathbf{G} = \begin{bmatrix} \mathbf{G}_s \\ \mathbf{g}_a \end{bmatrix},$$

where $\mathbf{G}_s$ is the generator of a strongly random-like code $\mathcal{C}_s$ of dimension K with a large minimum distance d_s, and where $\mathbf{g}_a$ is the generator matrix of a linear anticode of dimension k and small maximum distance d_a, with $k \ll K$ and $d_a \ll d_s$ [9]. The anticode is assumed to have a non-zero minimum distance d. Geometrically, we may think of shifted versions of the anticode, each centred on one of the 2^K words of $\mathcal{C}_s$.

At a small enough signal-to-noise ratio, decoding cannot separate the individual words of the shifted anticodes, although the wide separation of their centres makes the decoding errorless as regards the shifted anticode (i.e., the word of $\mathcal{C}_s$) identification. Of course, this is a definitely bad code according to the conventional word error rate criterion, but all the words of a given shifted anticode are close together so that a small BER results if d_a/n is small enough. This model will be helpful to us when dealing with pseudo-random convolutional codes and attempting to interpret the performance of turbo codes.

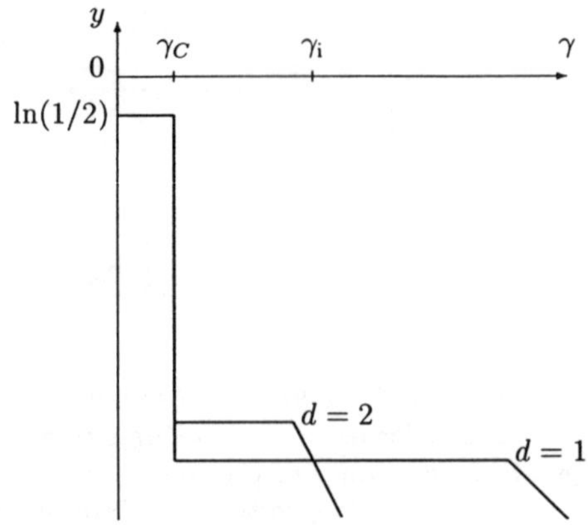

Figure 2: Broken line representation of the union upper bound on the bit error rate of the model of Sect. 3.3, as in Fig. 1 but for two values $d = 1$ and $d = 2$ of the minimum distance d. The other parameters are the same namely, $n = 1,024$, $R = 1/2$ and $A_d = 1$. It may be noticed that the two curves intersect at $\gamma_i = (1/2)\ln(n)$ so the code performs better for $d = 1$ than for $d = 2$ in the signal-to-noise interval $\gamma_C \leq \gamma \leq \gamma_i$.

4 Using convolutional coding for the design of weakly random-like codes

4.1 Recursive vs non-recursive convolutional codes

A striking feature of turbo codes is the fact they combine *recursive* convolutional codes. Recursivity is necessary for achieving the outstanding performance which has been reported by Berrou et al. [4]. However, according to the conventional theory of convolutional codes, there is an equivalence between the recursive and non-recursive convolutional codes as regards the set of generated sequences. Recursivity was deemed of no interest and Forney wrote: "... in fact, feedback buys nothing" [10]. We propose to revisit this problem and ask the question: how can we use convolutional coding in order to generate random-like codes?

First of all, we shall notice that the infinitely extending sequences which are assumed by the theory (denoted by formal series in the delay operator D and represented by diagrams like a tree or a trellis) do not belong to the real world. All sequences in engineering are finite, although maybe very long and/or of indeterminate length. We shall thus think of convolutional coding as a means for generating arbitrarily long, but finite, coded sequences in a much simpler manner than large-block coding and which do not explicitely involve the sequence length. The performance obtained will be almost completely determined by the parameters of the convolutional code and thus independent of the actual length. We shall thus assume that any message has a finite length of ℓ input symbols, so the corresponding coded output comprises ℓ/R symbols, where R is the code rate.

Now, if we search for convolutionally generated random-like codes, we may *a priori discard non-recursive encoding* because it associates low-weight coded sequences to low-weight input sequences. A necessary condition for obtaining random-like encoding is that the weight of coded sequences be as far as possible independent of that of input sequences. On the contrary, sequences with large weight (depending on the actual block length, but carved out from infinite length and infinite weight sequences) are generated by a recursive encoder in response to low-weight inputs. We may think of these codes as *anticatastrophic* since they associate infinite weight output sequences to finite weight input sequences (the main property of catastrophic codes is expressed by swaping in this sentence the words "input" and "output").

¿From now on we shall thus consider only recursive codes. Let us look at a rate-1/2 systematic recursive encoder of generator matrix

$$\mathbf{G}_{\text{rec}} = [1 \quad \text{N}(D)/\text{P}(D)], \tag{3}$$

where $\text{P}(D)$ is a polynomial of degree ν which specifies the recursive part of the encoder and $\text{N}(D)$ is a polynomial of degree at most equal to ν which corresponds to its non-recursive part. Since we are interested in information sequences of finite length ℓ, we may write the set of all coded sequences as

$$\mathcal{E} \triangleq \{m(D)\mathbf{G}_{\text{rec}} = [m(D) \quad m(D)\text{N}(D)/\text{P}(D)],\ 0 \leq \deg[m(D)] < \ell\}. \tag{4}$$

Some of the polynomials $m(D)$ are multiple of the denominator $\text{P}(D)$ i.e.:

$$m(D) = m'(D)\text{P}(D)$$

and we have obviously $0 \leq \deg[m'(D)] < \ell - \nu$. We may thus write the corresponding set of sequences as:

$$\mathcal{E}' \triangleq \{[m'(D)\text{P}(D) \quad m'(D)\text{N}(D)],\ 0 \leq \deg[m'(D)] < \ell - \nu\}.$$

But the matrix

$$\mathbf{G}_{\text{nrec}} \triangleq [\text{P}(D) \quad \text{N}(D)] \tag{5}$$

is the generator of the non-recursive code equivalent to the systematic recursive code of generator matrix $\mathbf{G}_{\text{rec}}$ given by (3). We may thus rewrite the set of sequences which are multiple of the denominator as:

$$\mathcal{E}' = \{m'(D)\mathbf{G}_{\text{nrec}},\ 0 \leq \deg[m(D)] < \ell - \nu\}. \tag{6}$$

This is a subset which contains only $2^{\ell-\nu}$ sequences among the 2^{ℓ} possible ones.

All the sequences which belong to $\mathcal{E}$ but not to $\mathcal{E}'$ have a length and a weight which indefinitely increase with ℓ so they do not contribute to decoding errors at a rate below the channel capacity. Only the fraction $\mathcal{E}'$ of all sequences, namely those generated by the equivalent non-recursive encoder, have small length and weight, but choosing ν large enough makes this fraction $2^{-\nu}$ very small. The situation resembles that of the model of Sect. 3.3 except that the equivalent non-recursive code $\mathcal{E}'$ plays here the same role as the anticode. Interestingly, certain results of simulation or analysis of turbo codes exhibit a variation of the bit error rate in terms of the signal-to-noise ratio which closely resembles the curves of Figs 1 and 2 [11, 12].

4.2 Pseudo-random recursive convolutional codes

We aim at designing encoders which generate a set of sequences as close as possible to random sequences (except for the minority subset $\mathcal{E}'$). If we take for $\mathrm{P}(D)$ in (3) a primitive polynomial, the encoder of such a code constitutes, in the absence of an input information sequence but assuming the encoder initial state to be different from 0, a generator of "maximum length sequences", often referred to as *pseudo-random*, so we shall refer to this encoder as pseudo-random, too. Fig. 3 represents an encoder of this type, with $\mathrm{P}(D) = 1 + D + D^3$ and $\mathrm{N}(D) = 1$.

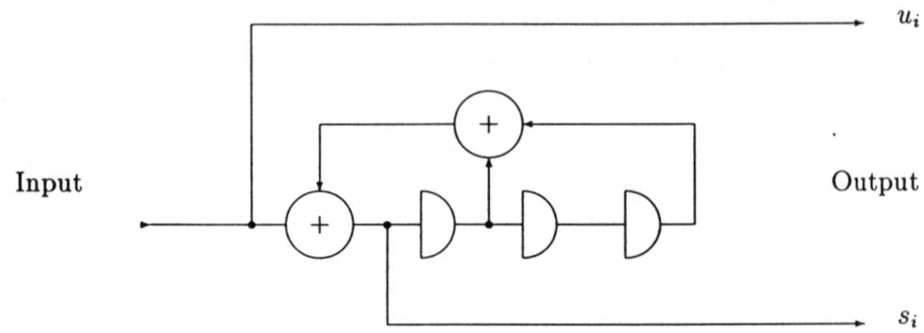

Figure 3: Example of a systematic recursive pseudo-random encoder for $\nu = 3$. Each point may assume one of two states denoted by 0 and 1. The D-shaped boxes represent delay operators one symbol duration long, and taken together they constitute a register. The sign + means modulo 2 addition.

The transition diagram of this encoder may be drawn as shown in Fig. 4. The closed path in heavy lines corresponds to the succession of states which is ran in the absence of input '1' when the initial state of the encoder is non-zero.

For lack of the exact weight distribution of a recursive convolutional code, an approximate estimation of it can be made. In the case of a pseudo-random (PRR) code i.e., which has a generator matrix of the form (3) where $\mathrm{P}(D)$ is a primitive polynomial of degree ν, there exist two cycles in the complete transition graph which are indefinetely ran in the absence of information symbols 1: the self loop at point a (which represents the 0 state) which has obviously 0 weight, and another cycle which connects all the other $2^\nu - 1$ points of the graph corresponding to the non-zero states. The cumulated weight of this latter cycle as drawn in Fig. 4 is positive and, for this reason, the weight of all paths which do not go back to point a tends to infinity with ℓ.

Let $\mathcal{S}^*$ denote the set of points which belong to this cycle. Let us consider one of its points, assumed to correspond to the encoder current state, and let us examine the transition to the next state. If the information symbol is 0, this transition leads to the next point of the cycle. If the information symbol is 1, the transition leads to

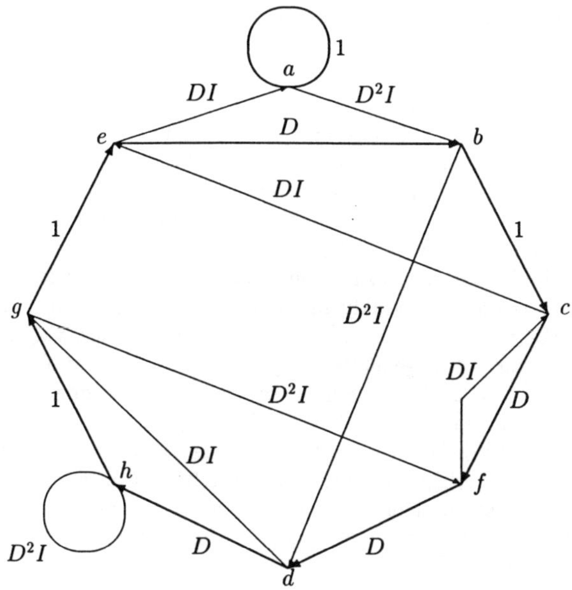

Figure 4: Transition graph of the encoder of Fig. 3. The states are denoted by $a, b, \ldots, h$, where a represents the all-0 state. The monomial which labels a branch is such that the exponent of D is its weight and that of I is the corresponding information bit. All the information bits are 0 along the closed path $\mathcal{S}^*$ drawn in heavy lines but its weight i.e., the sum of the exponents of D for all its branches, is non-zero.

another point of $\mathcal{S}^*$, except in the sole case where the starting point represents the state corresponding to a single 1 in the last register position; then, the transition leads to point a i.e., to the only point of the whole graph which does not belong to the cycle $\mathcal{S}^*$. If ν is as large as to make the probability of this case negligible, then one may describe the encoder operation as a random walk in the transition graph reduced to the points of $\mathcal{S}^*$. The steady-state probabilities of all states are equal. The weight associated with each transition is enumerated by the polynomial $[(x+1)/2]^2$, since each branch may have the weights 0, 1 and 2 with the probabilities 1/4, 1/2 and 1/4, respectively. The weight distribution associated with a sequence of ℓ transitions results from developing $W(x) = [(x+1)/2]^{2\ell}$: the frequency of a path of weight w is the coefficient of x^w in this series namely, $\binom{2\ell}{w}/4^\ell$. Therefore, within the validity limits of the approximation, the weight distribution is binomial i.e., the same as provided by an independent equiprobable random choice. The mean of this distribution is ℓ and its variance is $\ell/2$. Notice that the approximation made neglected the smallest weights i.e., the only ones which are responsible for errors, which are located in the subset $\mathcal{E}'$. Its validity is thus limited to large values of ν.

In the case of a non-recursive code, the average weight would be almost independent of ℓ and much smaller, in fact of the same order as ν. In effect, no cycle of positive weight is ran when the sequence of information bits is made only of zeros. As soon as the information symbol is 0, the path goes back to point a by the shortest way, and this point is reached after at most ν successive 0 information symbols. This results in an unequal steady-state probability of the points, those being close to a, whether in the way from, or back to it, being more frequently visited than the other ones.

One may thus think that pseudo-random recursive convolutional coding suffices to solve the problem of designing a weakly random-like code. By increasing the parameter ν which measures the memory of the encoder, we both decrease the fraction η of non-random-like words in the set of generated sequences and improve the approximation of the weight distribution of the other sequences to that of random coding. There are however difficulties in increasing ν. First of all, one thus exponentially increases the decoding complexity if the decoding algorithm relies on the trellis diagram, as the Viterbi algorithm or the MAP algorithm of Bahl et al. We believe however that this difficulty may be alleviated if we use an algorithm the complexity of which does not exponentially increase with ν e.g., non-exhaustive replication decoding, the non-optimality of which is compensated for by iteration [13].

There is also a less obvious, but fundamental, difficulty. Decoding by the Viterbi algorithm, for instance, consists of comparing metrics of paths as soon as they merge in the trellis diagram. However, except for the sequences which belong to the minority subset $\mathcal{E}'$, merging will the all-zero path does not occur. Using the Viterbi algorithm actually limits the search for sequences in $\mathcal{E}'$, which are very unlikely in the recursive case. In other words, this algorithm is well fitted only to the equivalent non-recursive code. A more appropriate decoding algorithm should take account of the fact that encoding is recursive, which only implies a difference in the probability distribution of the encoded sequences.

Instead of attempting to decode a large-memory pseudo-random convolutional code which undergoes the above difficulties, we may think of combining such codes according to the turbo code scheme. Then, decoding will reduce to that of separate small-memory codes. Moreover, using a properly designed interleaver enables decreasing the low-weight distribution tail, as will now be shown.

5 Combining convolutional codes according to the turbo code scheme

5.1 Concatenated and interleaved convolutional coding

Let us go back to the simple models of Sect. 3.3 and especially to that involving an anticode (as simpler for the analysis than the more realistic minority subset $\mathcal{E}'$). We have seen that decoding at a small signal-to-noise ratio, although larger than γ_C, cannot separate the words of a given shifted anticode, although its centre can be determined without error.

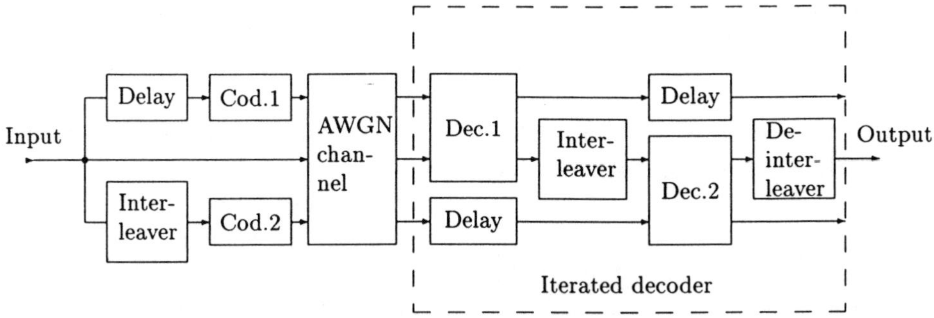

Figure 5: Encoder and decoder of a pseudo-random turbo code of rate 1/3. The boxes labelled Cod.1 and Cod.2 generate the redundancy bits corresponding to the input sequence according to a pseudo-random recursive code of rate 1/2. The box labelled "Iterated decoder" represents an elementary decoder with an output of same form as its input. It may thus be iterated, and it should actually be so a sufficient number of times in order to provide useful performance.

Let us now assume that the transmitted sequence, besides belonging to such a shifted anticode, also belongs to another completely independent subset. This may be achieved by a similar encoding of the same information sequence, but after interleaving which may be interpreted as replacing it by some random permutation. This scheme is that of turbo codes, as depicted for example in Fig. 5. Then, decoding will result in locating the correct sequence in both the chosen anticodes associated with the two encodings. It will thus be found in the intersection of both, which will contain much less many words. The word error rate will thus be widely improved.

We may interpret this result in terms of weight distribution and show that the low-weight tail of the obtained distribution has actually been reduced. Let us recall that we deal with weight distributions as with probability distributions. On the other hand, we may think of an interleaver as effecting a *random permutation* of the input data. Inasmuch as coding acts at random, each encoder independently associates a random redundancy sequence to each permutation of the input sequence. The overall redundancy sequence thus generated has therefore a weight which is the sum of independent random weights, each with a weakly random-like distribution i.e., close to that of random coding except maybe in its tails due to the minority subset of sequences $\mathcal{E}'$. The normalized weight distribution of the overall redundancy sequence is thus the *convolution product* of those of the constituent codes [14]. It is close to the average weight distribution of random coding in its median part (as it already was for the constituent codes), but the convolution product almost squares the distribution in its low-weight tail, where it is much larger than that of the random distribution.

Moreover, the rate of this code may be increased to 1/2 by puncturing, which consists here of alternately taking as output one of the two redundancy bits from the encoders. Inasmuch as the redundancy sequences have a (pseudo-) random distribution, the sequence resulting from their puncturing keeps this property.

5.2 A multiple-component turbo code as a general means for designing random-like codes

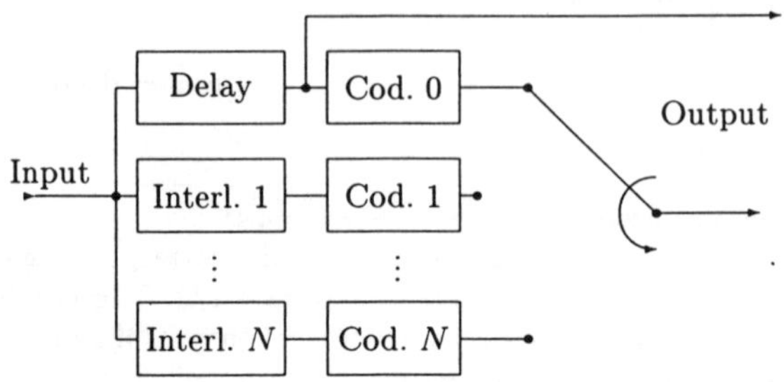

Figure 6: Encoder of a pseudo-random turbo code of rate 1/2. The boxes labelled "Cod. 0, Cod. 1, ..., Cod. N" generate the redundancy sequences associated with the input sequences according to a pseudo-random code of rate 1/2. The boxes labelled "Interl. 1, ..., Interl. N" represent interleavers, the delay of which is compensated for in the box labelled "Delay". Periodic puncturing keeps the rate to its initial value 1/2.

We may extend the punctured turbo code scheme according to Fig. 6. There, we combine an arbitrary number, $N+1$, of rate-1/2 PRR codes and periodic puncturing of the redundancy sequences increases the overall rate to 1/2. The weight distribution of the overall redundancy sequence generated, puncturing being provisionally ignored, is now the convolution product of the weight distributions of the $N+1$ redundancy sequences. Due to the central limit theorem, it approaches a Gaussian distribution of mean $(N+1)\ell/2$ when the message length ℓ becomes very large. This is the well known approximation of the weight distribution of a Bernoulli sequence of parameter 1/2. Increasing N will make the distribution closer and closer to that of random coding, which will result in a fast decrease of the BER as the number of constituent codes increases. Moreover, inasmuch as the redundancy sequences thus generated mimic random sequences, their periodic puncturing keeps a random-like weight distribution.

The interleavers are intended in this scheme to feed the encoders with permutations of the information sequence such that the corresponding generated redundancy sequences can be assumed independent, and the above conclusion holds only inasmuch as this assumption is true. This excludes a number of interleaver schemes which result in too regular permutations, for instance cyclic shifts.

Of course, using recursive pseudo-random convolutional codes as components in the turbo code scheme will result in a weight distribution very close to that of random coding for small values of the number $N+1$ of constituent codes, even for $N=1$ as in the previous section, since the initial weight distribution of each of these codes is itself close to that of random coding except maybe in the distribution tails. However, the central limit theorem results in an asymptotically random weight distribution regardless of the initial one as N tends to infinity (provided the component codes satisfy weak conditions such as being balanced), so the turbo code scheme with many constituent codes may be considered as *intrinsically random-like.*

6 Conclusion

Designing codes so as to mimic random coding does not seem a very difficult task if, instead of strongly random-like codes, we look for weakly random-like ones. The analysis of pseudo-random convolutional codes outlined above shows they actually belong to this family. The difficulty of their decoding especially if their memory is large, as well as their rather poor closeness to random coding if it is not, are widely improved by the use of schemes combining concatenation and interleaving. The turbo code scheme thus appears as an outstanding example of a simple means to generate weakly random-like codes with a specified bit error rate at an information rate close to the channel capacity.

Appendices

A Strongly random-like codes meet the Gilbert-Varshamov bound

Let us assume that we consider a strongly random-like code i.e., such that each A_w in its weight enumerator is close to the average number of words of weight w obtained in the average by random coding namely, for an (n, k) binary code, $R_w = 2^{-(n-k)}\binom{n}{w}$. For instance, A_w may be the nearest integer to this average number. One should then have $A_w = 0$ if $2^{-(n-k)}\binom{n}{w} < \lambda$, where λ is some positive constant e.g., 1/2, for an approximation to the nearest integer. The first non-zero coefficient A_d thus corresponds to the smallest weight d such that $2^{-(n-k)}\binom{n}{d} \geq \lambda$. Assuming that n and d are as large as to enable using the Stirling approximation, d is the smallest integer such that

$$\frac{n^{n+1/2}}{\sqrt{2\pi}} d^{d+1/2}(n-d)^{n-d+1/2} \geq \lambda 2^{n-k}.$$

Taking logarithms to the base 2 and dividing by n results in

$$\begin{aligned}(1+1/2n)\log_2 n - (1/2n)\log_2(2\pi) - (d/n+1/2n)\log_2 d \\ - (1-d/n+1/2n)\log_2(n-d) \geq 1-k/n+(1/n)\log_2\lambda.\end{aligned}$$

Letting $d/n = \delta$, $k/n = R$ and neglecting the terms in $1/n$ results in the inequality:

$$\mathcal{H}_2(\delta) \geq 1-R, \tag{7}$$

where $\mathcal{H}_2(x) \triangleq -x\log_2(x) - (1-x)\log_2(1-x)$ is the entropy of a binary random variable one of the outcomes of which occurs with probability x, expressed in binary unit. Notice that the parameter λ which specifies the approximation no longer appears in (7). This inequality turns out to be the Gilbert-Varshamov bound. We may define the Gilbert-Varshamov distance of an (n, Rn) code as $d_{\text{GV}} \triangleq n\mathcal{H}_2^{-1}(1-R)$ where $y = \mathcal{H}_2^{-1}(x)$, $0 < y \leq 1/2$, is such that $x = \mathcal{H}_2(y)$.

For a q-ary linear code, the same argument results in the Gilbert-Varshamov bound in the form

$$\mathcal{H}_q(\delta) + \delta\log_q(q-1) \geq 1 - R\,, \tag{8}$$

with $\mathcal{H}_q(x) \triangleq -x\log_q(x) - (1-x)\log_q(1-x)$. Now $d_{\text{GV}} = n\delta_0$, where δ_0 is the smallest positive solution of the implicit equation which results from (8) by substituting $=$ for $\geq$.

B Noise geometry in signal space

Assuming their band limited to B and applying the sampling theorem, Shannon represented signals considered on a certain time interval T by points in the Euclidean space of dimension $n = 2BT$ [1]. Then, the squared Euclidean norm represents an energy and the noise, assumed to be additive Gaussian and white with one-sided spectral density N_0, has independent components of same variance $N_0/2$.

The asymptotic behaviour of the noise as the space dimension n approaches infinity is described by the so-called "sphere hardening phenomenon" [1]. Quoting Shannon, the noise spheres "can be thought of roughly as sharply defined billiard balls". The norm of the noise vector $r = \sqrt{\sum_{i=1}^{n} n_i^2}$, where n_i is the i-th component of the noise vector, of zero mean and variance $N_0/2$, tends as n approaches infinity to be quasi-deterministic. Its mean is

$$\overline{r} = \sqrt{N_0}\Gamma(n/2 + 1/2)/\Gamma(n/2),$$

where $\Gamma(.)$ is the Eulerian factorial function. It behaves as $\sqrt{nN_0/2}$ as n approaches infinity, but the variance of r

$$\text{var}(r) = \text{E}[(r-\overline{r})^2] = N_0\{n/2 - [\Gamma(n/2+1/2)/\Gamma(n/2)]^2\} \tag{9}$$

approaches zero as $n \to \infty$, although the mean $\overline{r}$ indefinitely increases. If we assume a constant signal power P and an increasingly large time interval T, with n proportional to T according to $n = 2BT$, then we may divide the distances by $\sqrt{T}$ and the squared Euclidean norm becomes proportional to a power. The mean radius of the noise spheres becomes close to the constant $\sqrt{N}$ where $N = N_0B$ is the noise power in B.

The meaning of the sphere hardening phenomenon is just a statistical compensation: some components are larger than average, some others are less, but due to their large number it is very unlikely that a significant majority of them be larger or smaller than the mean. A noise sphere is indeed a sharply defined ball *as an n-dimensional object* but not if we consider some of its properties in low-dimensional space. For instance, the noise component in any straight line passing by the sphere centre is just an ordinary Gaussian random variable of variance $N_0/2T$ due to the isotropic distribution of the noise, so its distribution is quite smooth. Therefore, the packing of spheres which ensures a separation of $d > \sqrt{N}$ between the noise spheres does not solve the problem of errorless communication at a rate close to the capacity. We may also notice that, if it were so, the density of sphere packings would approach 1 as n approaches infinity. Bounds are known, however, which show that it actually tends to 0. For transmitting at a rate close to the channel capacity, the sphere centres should not only be separated by a sufficient minimum distance, but the centres the closest to a given one should be as scarce as to make the probability of the noise vector to go towards a neighbour very small.

C Union bound analysis of the bit error rate of the crude model of Section 3.3

The union bound on the word error rate (WER) reads:

$$P_{\mathrm{WER}} \leq \sum_{w} A_w P(C_w|C_0), \tag{10}$$

where A_w is the number of words of weight w and $P(C_w|C_0)$ is the probability of receiving a particular codeword C_w of weight w when the all-zero word C_0 is transmitted. For the additive Gaussian white noise channel, we have:

$$P(C_w|C_0) = \frac{1}{2}\mathrm{erfc}(\sqrt{wE_s/N_0}), \tag{11}$$

where E_s is the received energy per channel symbol and N_0 the noise one-sided spectral density. The "complementary error function" is defined as

$$\mathrm{erfc}(x) \triangleq \frac{2}{\sqrt{\pi}} \int_x^{\infty} \exp(-t^2)\mathrm{d}t.$$

The bound on the bit error rate (BER) corresponding to (10) is:

$$P_{\mathrm{BER}} \leq \sum_{w} \frac{w}{n} A_w P(C_w|C_0), \tag{12}$$

assuming without loss of generality that an information bit is as likely to be in error as any other bit in the word.

A rough but very convenient approximation of the complementary error function is well known to be:

$$\mathrm{erfc}(x) \approx \exp(-x^2).$$

Taking account of (11) and using this approximation, we may rewrite (12) as:

$$P_{\mathrm{BER}} \lesssim \sum_{w} \exp[a_w - \gamma w + \ln(w) - \ln(n)], \tag{13}$$

where $a_w \triangleq \ln(A_w/2)$ and $\gamma \triangleq E_s/N_0$. Now, we represent each term of the right hand side of (13) by a line of equation

$$y_w \triangleq a_w - \gamma w + \ln(w) - \ln(n) = b_w - \gamma w\,. \tag{14}$$

An obvious upper bound on this term is also w/n, so we may represent the upper bound on the bit error probability associated with some term of weight w as [5]:

$$y = \min[y_w, \ln(w/n)] = \min[b_w - \gamma w, \ln(w/n)], \tag{15}$$

where b_w is defined by the second equality in (14). In the crude model of Sect. 3.3, we may think of the set of large weight words as achieving the channel capacity C so its performance may be represented by an almost vertical line at γ_C if n is large. To take account of the single term of weight d, which dominates the performance for $\gamma > \gamma_C$, we have the single curve given by (15) for $w = d$, as in Fig. 1.

References

[1] C.E. Shannon, "Communication in the presence of noise," *Proc. IRE*, pp. 10–21, Jan. 1949.

[2] G. Battail, "Construction explicite de bons codes longs," *Annales Télécommunic.*, Vol. 44, No. 7–8, pp. 392–404, Jul.-Aug. 1989.

[3] G. Battail, H. Magalhães de Oliveira, and W. Zhang, "Codage déterministe imitant le codage aléatoire pour le canal à bruit gaussien additif," *Annales Télécommunic.*, Vol. 47, No. 9–10, pp. 443–447, Sept.-Oct. 1992.

[4] C. Berrou, A. Glavieux, and P. Thitimajshima, "Near Shannon Limit Error-Correcting Coding and Decoding : Turbo-Codes," *Proc. of ICC'93*, Geneva, Switzerland, pp. 1064–1070, 23-26 May 1993.

[5] G. Battail, "On random-like codes," *Proc. 4-th Canadian Workshop on Information Theory*, Lac Delage, Québec, 28–31 May 1995. Lecture Notes in Computer Science No. 1133, pp. 76–94, Springer, 1996.

[6] G. Battail, "Pseudo-random recursive convolutional coding for near-capacity performance," *2-nd International Symposium on Communication Theory and Applications,* Ambleside, U.K., 11–16 July, 1993; B. Honary, M. Darnell and P. Farrell Eds., *Communications and Signal Processing Series: Communications Theory and Applications II*, pp. 54–65, HW Communications Ltd., 1994.

[7] G. Battail, C. Berrou, and A. Glavieux, "Pseudo-random recursive convolutional coding for near-capacity performance," *Proc. GLOBECOM'93, Communication Theory Mini-Conference*, Vol. 4, pp. 23–27, Houston, U.S.A., Nov. 29-Dec. 2, 1993.

[8] G. Battail, "Can we implement random coding?," *4-th IMA Conference on Cryptography and Coding,* Cirencester, U.K., 13–15 Dec., 1993; *Codes and Cyphers*, P.G. Farrell, Ed., pp. 1–15, Formara Ltd, Southend-on-Sea, U.K., 1995.

[9] P.G. Farrell, "Linear binary anticodes," *Electronics Letters,* Vol. 6, pp. 419–421, 1970.

[10] G.D. Forney Jr, "Convolutional Codes I: Algebraic Structure," *IEEE Trans. Inf. Th.*, Vol. IT-16, No. 6, pp. 720–738, Nov. 1970.

[11] P. Robertson, "Illuminating the structure of code and decoder of parallel concatenated recursive systematic (turbo) codes," *Proc. GLOBECOM'94*, pp. 1298–1303, San Francisco, U.S.A., Nov. 27– Dec. 1, 1994.

[12] S. Benedetto, G. Montorsi, "Unveiling turbo-codes: some results on parallel concatenated coding schemes," *IEEE Trans. Inf. Th.*, Vol. 42, No. 5, pp. 409–429, Mar. 1996.

[13] G. Battail, "A conceptual framework for understanding turbo codes," submitted to *IEEE Journal on Selected Areas in Comm.*

[14] G. Battail, "Pseudo-random turbo-codes," *Proc. Int. Symp. on Signals, Systems and Electronics*, San Francisco, U.S.A., pp. 419–422, Oct. 25–27, 1995.

4. Concatenated Signal Modulation Techniques for RS Coded Modulation

B. Honary[(1)]
R. Edwards[(1)]
S. Farquhar[(2)]
A. Noble[(2)]

[(1)] Lancaster Communications Research Centre, Lancaster University, Lancaster. LA1 4YR. UK
[(2)] DRA, St. Andrews Road, Malvern, Worcs. WR14 3PS. UK

Abstract: A novel approach to designing signal sets for block coded modulation is introduced. A signal set is expanded followed by a more moderate reduction in the ensuing constellation size. Successive signal points are concatenated to produce greater dimensional signal points; the bandwidth expansion being compensated for by a large increase in minimum squared Euclidean distance. A Single Parity Check code is then used to partition the expanded signal constellation to produce a final signal constellation with greater minimum Euclidean distance, but only half the number of signal points. These modulation schemes then are combined with RS codes to form coded modulation schemes offering up to 6dB gain, asymptotically, for bandwidth efficient schemes over an additive white Gaussian noise (AWGN) channel. Similar gains are shown to be attainable over a Rayleigh Fading Channel. We show that by concatenating more complex modulation schemes even gains in excess of 10dB asymptotically are achievable over both the AWGN and Rayleigh Channels.

1. Introduction

Coded modulation [1, 2] combines the coding and modulation processes to improve error correction capabilities. Both the coding and the modulation are chosen carefully to maximise the minimum squared Euclidean distance (MSED). There are two approaches to maximising the Euclidean distance of the scheme. The first approach is to use inner and outer codes [3] with a set-partitioned signal constellation: the more powerful codes being used to encode the partitions with smaller Euclidean distances; the less powerful codes being used to encode the partitions with larger Euclidean distances. The second approach is to map codeword symbols directly to members of the signal set [4]. Performance of coding schemes using this approach may be further increased by concatenating signal points to increase the MSED of the coded modulation. The MSED of a scheme produced using a mapping process will be the product of the MSED of the signal constellation and the minimum distance of the code. The performance of such coded modulation schemes is dependent on the use of powerful codes and efficiently paired modulation schemes. Reed-Solomon codes are multilevel cyclic codes, with powerful error correcting capabilities, and provide the large distances required for an efficient coded modulation scheme. By concatenating phase-varying modulation signals [5,6,7] to produce an expanded signal constellation, we can design more powerful coded modulation schemes with greater minimum squared Euclidean distance; allowing us to construct bandwidth efficient coded modulation schemes. A bandwidth efficient coded modulation scheme being such that error performance is improved without a corresponding sacrifice in data rate. By concatenating a four-dimensional 2FSK/MPSK modulation scheme [8,9] that both varies in phase and frequency, we substantially increase the MSED and hence further improve performance.

We first describe the design of a suitable concatenated MPSK signal set. A two-dimensional MPSK set is concatenated N times to produce a 2N-dimensional signal set. If the original signal set has been labelled using a gray code, we can then use a Single Parity Check (SPC) to halve the number of signal points in the constellation, whilst doubling the minimum squared Euclidean distance.

We then describe how these signal constellations may be combined with RS codes to produce coded modulation schemes, with some bandwidth efficient schemes introduced.

We finally present a range of simulation results and some conclusions.

2. Signal Set Design

We concatenate N MPSK signal points [5,6], where N is an integer greater than one to produce a single 2N-dimensional signal point, and a new signal set of M^N signal points is derived. That is for every modulation symbol that would originally have been transmitted, N symbols are now transmitted instead. By transmitting a greater number of modulation symbols, a greater separation may be achieved between signal points for the same overall signal constellation size. The transmitted signal, s(t), may be expressed as a combination of 2N basis functions, such that

$$s_j(t) = c_1\psi_1(t) + c_2\psi_2(t) + \cdots + c_{2N}\psi_{2N}(t) \qquad \text{for } 0<j<M^N \tag{1}$$

The signal may thus be represented by a 2N-dimensional co-ordinate $(c_1, c_2, \ldots, c_{2N})$, and the basis functions are given by

$$\psi_{2i+1} = \sqrt{\frac{2}{T}}\cos\omega_c t, \text{ where } iT < t < (i+1)T, \text{ for } 0<i<N$$
$$\psi_{2i+2} = \sqrt{\frac{2}{T}}\sin\omega_c t, \text{ where } iT < t < (i+1)T, \text{ for } 0<i<N \tag{2}$$

where T is the modulation interval of each two-dimensional signal and NT the modulation interval of each 2N-dimensional signal.

Example 1. 2*4PSK Signal Set. Two 4PSK signal sets are concatenated, thus N=2, with M=4. The new signal set consists of 4^2=16 signal points. Using eqns.1,2, the 2N-dimensional co-ordinates (c_0, c_1, c_2, c_3) are shown below for each s_i, and are illustrated graphically in Fig. 1:

$$\begin{array}{llll} s_0 = (1,0,1,0) & s_1 = (1,0,0,1) & s_2 = (1,0,-1,0) & s_3 = (1,0,0,-1) \\ s_4 = (0,1,1,0) & s_5 = (0,1,0,1) & s_6 = (0,1,-1,0) & s_7 = (0,1,0,-1) \\ s_8 = (-1,0,1,0) & s_9 = (-1,0,0,1) & s_{10} = (-1,0,-1,0) & s_{11} = (-1,0,0,-1) \\ s_{14} = (0,-1,1,0) & s_{13} = (0,-1,0,1) & s_{14} = (0,-1,-1,0) & s_{15} = (0,-1,0,-1) \end{array} \quad (3)$$

These signals may be represented as the combination of two 4PSK signal points, the first signal point valid for $0<t<T$, and the second valid for $T<t<2T$.
Each of the M^N signal points may be represented by an $N*\log_2 M$ binary label. If each of the N 2-dimensional signal constellations is labelled using a gray code mapping, as shown in Fig. 2, then each binary label is the concatenation of the N corresponding M-bit gray coded labels.

2.1 Concatenated 2FSK/MPSK

A 4N dimensional signal constellation is designed by concatenating two four-dimensional 2FSK/MPSK signal constellations [8]. Each modulation signal in the new constellation exists over a period of 2T. Each new signal point, $s_i(t)$, may be considered as a linear combination of eight basis functions, $\psi_j(t)$, such that

$$s_i(t) = c_1\psi_1(t) + c_2\psi_2(t) + \cdots + c_{4N}\psi_{4N}(t) \quad (4)$$

where c_j, j=1,2,…,8 are the vectorial coefficients. The basis functions are chosen to be

$$\psi_{4i+1} = \sqrt{\frac{2}{T}}\cos\left[\omega_c t + \frac{\pi t}{T}\right] \quad \text{where } iT \le t < (i+1)T, \text{ for } 0<i<N$$

$$\psi_{4i+2} = \sqrt{\frac{2}{T}}\sin\left[\omega_c t + \frac{\pi t}{T}\right] \quad \text{where } iT \le t < (i+1)T, \text{ for } 0<i<N$$

$$\psi_{4i+3} = \sqrt{\frac{2}{T}}\cos\left[\omega_c t - \frac{\pi t}{T}\right] \quad \text{where } iT \le t < (i+1)T, \text{ for } 0<i<N$$

$$\psi_{4i+4} = \sqrt{\frac{2}{T}}\sin\left[\omega_c t - \frac{\pi t}{T}\right] \quad \text{where } iT \le t < (i+1)T, \text{ for } 0<i<N \quad (5)$$

where T is the modulation interval of each two-dimensional signal and NT the modulation interval of each 2N-dimensional signal.

Example 2. ***2*2FSK/8PSK Signal Set.*** In this example we shall describe the construction of the 2*2FSK/8PSK signal constellation. This is an eight-dimensional signal constellation with 256 signal points with the basis functions

$$\psi_1 = \sqrt{\frac{2}{T}}\cos\left[\omega_c t + \frac{\pi t}{T}\right] \quad \text{for } 0 \le t < T$$

$$\psi_2 = \sqrt{\frac{2}{T}}\sin\left[\omega_c t + \frac{\pi t}{T}\right] \quad \text{for } 0 \le t < T$$

$$\psi_3 = \sqrt{\frac{2}{T}}\cos\left[\omega_c t - \frac{\pi t}{T}\right] \quad \text{for } 0 \le t < T$$

$$\psi_4 = \sqrt{\frac{2}{T}}\sin\left[\omega_c t - \frac{\pi t}{T}\right] \quad \text{for } 0 \le t < T$$

$$\psi_5 = \sqrt{\frac{2}{T}}\cos\left[\omega_c t + \frac{\pi t}{T}\right] \quad \text{for } T \le t < 2T$$

$$\psi_6 = \sqrt{\frac{2}{T}}\sin\left[\omega_c t + \frac{\pi t}{T}\right] \quad \text{for } T \le t < 2T$$

$$\psi_7 = \sqrt{\frac{2}{T}}\cos\left[\omega_c t - \frac{\pi t}{T}\right] \quad \text{for } T \le t < 2T$$

$$\psi_8 = \sqrt{\frac{2}{T}}\sin\left[\omega_c t - \frac{\pi t}{T}\right] \quad \text{for } T \le t < 2T \qquad (6)$$

Each coefficient is chosen such that $c_j \in \{-1, -\frac{1}{\sqrt{2}}, 0, \frac{1}{\sqrt{2}}, 1,\}$ and that

$$\sum_{j=1}^{4} c_j^2 = \sum_{j=5}^{8} c_j^2 = 1 \qquad (7)$$

similarly,

$$\sum_{j=1}^{2} c_j^2 \neq \sum_{j=3}^{4} c_j^2 \text{ and } \sum_{j=5}^{6} c_j^2 \neq \sum_{j=7}^{8} c_j^2 \text{ where } \sum_{j=1}^{2} c_j^2, \sum_{j=3}^{4} c_j^2, \sum_{j=5}^{6} c_j^2, \sum_{j=7}^{8} c_j^2 \in \{0,1\}. \qquad (8)$$

Each of the 256 signal points may be represented by an eight bit binary label. This label is formed by concatenating two four bit labels from each of two gray coded 2FSK/8PSK signal points, as illustrated in Fig 3.

2.2 Set Partitioning

The signal set is partitioned using a Single Parity Check (SPC) code. Since each signal point may be represented by a binary label, the constellation may be partitioned into two equally sized groups: those where the label consist of an even number of ones; and those whose label contains an odd number of ones. Thus, the last bit in the label represents either an even or an odd parity check. Since the labels are of length $N*\log_2M$, then a $(N*\log_2M, (N*\log_2M)-1)$ SPC code is used. This divides the M^N-dimensional signal points into a smaller set consisting of half the original number of signal points. If an even parity check has been used, then an odd parity check will obviously identify the unused signal points. Since a gray code is used to produce the labels, then the minimum distance of each partition is as follows

$$d^2_{partition} = \min\,[2*\delta_0^2, \delta_1^2] \quad (9)$$

where δ_0^2, δ_1^2 correspond to a single MPSK signal set. The rate of the new modulation scheme, r, is

$$r = (N\log_2M - 1)/N \quad \text{bits/T} \quad (10)$$

Example 3. 2*4PSK with (4,3) SPC Code. Continuing Example 1; the 2*4PSK signal set is reduced to an eight member signal set using either an odd or even SPC code. The new information rate and minimum squared Euclidean distance are, respectively: $r = 3/2 = 1.5$; $d^2_{part} = \min[2*2,4] = 4$.

3. RS Coded Modulation

Combined Coding and Modulation (CCM) schemes may be formed by combining an (n,k,d) RS code over GF(q) with a multidimensional concatenated modulation scheme. Each Galois Field symbol is mapped to a unique signal point in the constellation [4]. The modulation scheme is chosen such that

$$q = \tfrac{1}{2}M^N \quad (11)$$

Each codeword consists of n symbols over GF(q), each of which may be represented by $\log_2 q$ binary bits. Using an SPC-partitioned concatenated modulation scheme each of the q Galois Field elements is mapped to one of the $\tfrac{1}{2}M^N$ 2N-dimensional signal points in the constellation as described previously. The information rate of the combined scheme is

$$r = \frac{mk}{Nn} \tag{12}$$

where m = $\log_2 q$, k is the number of information symbols over GF(q) per RS codeword, N is the number of concatenated signal points, and n is the number of GF(q) symbols per RS codeword.
The MSED of the RS-coded modulation scheme, d^2_{coded}, is given by

$$d^2_{coded} = d^* d^2_{partition} \tag{13}$$

where d is the minimum distance of the RS code, and d^2_{part} is the MSED of the SPC-partitioned modulation scheme. The rate and MSED will later be used to estimate the performance of the code.

Example 4. 2*4PSK/(4,3) SPC code with RS codes over GF(8). The eight Galois Field symbols over GF(8) are mapped to signal points from the 2*4PSK signal constellation using an even (4,3) parity check code, as shown in Table 1. Two coded modulation schemes may now be constructed over GF(8); based on the (7,3) and (7,5) RS codes. The scheme based on the (7,3) RS-coded modulation scheme has a rate, r = (3*3)/(2*7) = 9/14 = 0.643 bits/T; MSED, d^2_{coded} = 5*4 = 20. Similarly, for the (7,5) RS code: r = 15/14 = 1.07 bits/T; d^2_{coded} = 12.

4. Performance

We shall now estimate the performance of these schemes using the asymptotic coding gain to give a bound on the performance of the coded modulation schemes. Using this bound we shall estimate the performance of some coded modulation schemes that are bandwidth efficient and show by simulation the gains that are practically attainable.

4.1 Asymptotic Coding Gain

The Asymptotic Coding Gain, g_∞, over uncoded modulation for the AWGN channel [10], is given by

$$g_\infty = 10\log_{10} \frac{d^2_{coded}}{4\sin^2 \frac{\pi}{2^r}} \tag{14}$$

As N increases, r → $\log_2 M$, and $d^2_{part} = 2\delta_0^2$, g_∞ thus becomes

$$g_\infty \xrightarrow[\lim N\to\infty]{} 10\log_{10}\frac{2\delta_0^2}{\delta_0^2} = 3.01\text{dB} \tag{15}$$

Two schemes of differing rates may be compared using a modified version of eqn. 14. This takes into account the bandwidth expansion required by the coded scheme over the uncoded scheme [11], as shown below;

$$g_\infty = 10\log_{10}\frac{r_{coded}}{r_{uncoded}} + 10\log_{10}\frac{d^2_{coded}}{d^2_{uncoded}} \tag{16}$$

Example 5. Performance of GF(8) RS Coded 2*4PSK/(4,3)SPC Modulation.
Continuing Example 4, for the coded modulation scheme based on the (7,3) RS Code, with rate r_{coded} = 0.643 bits/T and d^2_{coded} = 20, the asymptotic coding gain over uncoded QPSK ($r_{uncoded}$=2bits/T, $d^2_{uncoded}$ = 2) using eqn. 16 is g_∞ = 5.07dB. Similarly, for the coded modulation scheme based on the (7,5) RS Code (rate r_{coded} = 1.07 bits/T and d^2_{coded} = 12, the asymptotic coding gain over uncoded QPSK ($r_{uncoded}$=2 bits/T, $d^2_{uncoded}$ = 2), g_∞ = 5.07dB. Both coded modulation schemes therefore predict the same performance over uncoded QPSK, although the (7,5) RS coded modulation scheme gives this performance at a greater information rate.

4.2 Bandwidth Efficient Schemes

A wide range of coded modulation schemes may be designed using this construction. We are particularly interested in schemes that are bandwidth efficient. These schemes are important since they offer improvements in both information rate and performance. That is, those that provide an increase in performance, without any reduction in error rate.
A group of RS-coded modulation schemes are introduced over GF(32) that are bandwidth efficient with respect to uncoded QPSK. RS codes over GF(32) are combined with 2*8PSK modulation, partitioned using a (6,5) parity check code. Table 2 shows the information rates and asymptotic coding gains over uncoded QPSK for those schemes that are bandwidth efficient. Gains of up to 6dB over uncoded QPSK are predicted. Similarly, Table 3 shows the information rates and gains for bandwidth efficient coded modulation schemes combining RS codes over GF(128) with a partitioned 2*2FSK/8PSK modulation scheme. We see that gains of up to 15dB are theoretically attainable. Due to the number of coded modulation schemes that are available based on RS codes over GF(128) we have produced a graph to enable us to choice the most appropriate scheme for our requirements. Fig. 4 shows asymptotic coding gain plotted versus information rate for all of the coded modulation schemes available. The graph shows those schemes that are bandwidth efficient with respect to uncoded QPSK. We can see that we can achieve gains from 15dB to 5dB with corresponding rates from 2 bits/T to 3.5 bits/T. We

can also see that going from a scheme with 2 bits/T to one with 3 bits/T we only sacrifice a small amount of performance, still achieving a potential gain of 12.5dB over uncoded QPSK.

It is possible to continue to produce a wide range of bandwidth efficient schemes beyond those already described, for example Table 4, shows that if by combining 2FSK/4PSK in place of 8PSK with RS codes over GF(32), gains of up to 11.50 dB may be achieved. Similarly we can also combine greater than two signal constellations to produce equally efficient bandwidth efficient schemes based on RS codes over larger Galois Fields.

4.3 Simulation Results

In this section we present simulation results for the modulation and coded-modulation schemes previously described.

Fig. 5 shows the simulation results for two concatenated and partitioned modulation schemes, with a 4PSK signal set concatenated two and three times respectively. It can be seen that these schemes achieve the predicted 2dB gain over uncoded QPSK.

Fig. 6 presents simulation results for two coded-modulation schemes using the (7,3) and (7,5) RS Codes over GF(8). These schemes have been implemented using soft decision trellis decoding [12,13]. Example 4 predicts gains of 5dB for these schemes and again the simulated performance approaches this bound. The (7,3) based scheme outperforms the (7,5) based scheme due to the lower complexity of its trellis.

Fig. 7 shows simulation results for the bandwidth efficient schemes over GF(32). Due to complexity considerations, we have used a hard decision decoding algorithm for implementation. Gains of 2-3dB are achieved commensurate with the asymptotic gains of 5-6dB predicted with soft decision decoding.

With Fig. 8 we present some selected results based on the 8-dimensional construction using RS codes over GF(128) and SPC partitioned concatenated 2FSK/8MPSK modulation. Due to the complexity of these codes, we have chosen to simulate them using a hard decision decoder, and so the performance does not quite achieve the bounds predicted earlier. However, gains of 6-7dB are practically achievable.

We have additionally applied a some schemes to a Rayleigh Fading Channel. We assume that the receiver performs coherent detection and that the channel phase shift is perfectly compensated for. An interleaver has been used to minimise the effect of burst errors. Fig. 9 shows that over a Fading Channel more substantial gains are possible with a 12dB gain at an error rate of 10^{-3} again using hard decision decoding.

5. Conclusions

An RS coded modulation technique using concatenated MPSK modulation and SPC codes is described. It is shown that the concatenated modulation schemes offer up to 3dB gain over uncoded MPSK modulation without additional coding.

We have presented the design of block coded modulation combining RS codes with these partitioned and concatenated modulation schemes and have shown by simulation that gains of 5-6dB over uncoded QPSK may be achieved.
We have also shown that by using a more complex construction based on 2FSK/MPSK larger gains are possible, with up to 15dB attainable over the AWGN channel, and 12dB achieved practically over the Rayleigh Fading Channel.
We have also shown that large gains are also attainable with a wide range of coded modulation schemes beyond those described in this paper.

6. References

[1] Ungerboeck, G., "Channel coding with multilevel/phase signals",.*IEEE Trans. I.T.*, 1982, vol. 28, No. 1, pp. 55-67

[2] Farell, P.G., Borelli, W., and Rashvand, H.W., "Convolutional Codes for Multi-level Modems", *Electronic Letters*, vol. 17, 1981, No. 9, pp. 331-333

[3] Calderbank, A.R., and Seshadri, N., "Multilevel Codes for Unequal Error Protection", *IEEE Trans. I.T.*, Vol. 39, No. 4, pp. 1234-1248, July 1993

[4] Jamali, S.H., and Le-Ngoc, L., "Performance Comparison of Different Decoding Strategies for a Bandwidth Efficient Block-Coded Scheme On Mobile Radio Channels", *IEEE Trans. Vehicular Tech.*, Vol. 41, No. 4, pp. 505-515, November 1992

[5] Asgari, V., and Shiva, M., "Channel Coding for VSAT Systems", *IEE Proc. Commun.,* Vol. 142, No. 5, pp. 333-339, October, 1995

[6] Jamali, S.H., and Le-Ngoc, L., "Bandwidth Efficient Communication via a Rayleigh Fading Channel Using RS Coded Multiphase Signalling", *IEEE Trans. Commun.*, Vol. 41, No. 11, pp. 1594-1597, November 1993

[7] Edwards, R, Honary, B., Farquhar, S., and Noble, A., "RS-Coded Modulation Employing Concatenated Signal Modulation", *Proc. 4th International Symposium on Comm. Theory and Applications,* p.383, July 1997

[8] Padovani, R., and Wolf, J.K., "Coded phase/frequency modulation", *IEEE Trans. Commun.*, 1986, vol. 34, No. 5, pp. 446-453

[9] Edwards, R, Honary, B., Farquhar, S., and Noble, A., "Advanced Bandwidth Efficient Block Coded Modulation Over Fading Channels", *Proceedings MILCOM 1997*, 2nd-5th November 1997

[10] Kschischang, F.R., de Buda, P.G., Pasupathy, S., "Block Coset Codes for M-ary Phase Shift Keying", *IEEE Journal Selected Areas Commun.,* Vol. 7, No. 6, pp. 900-913, August 1989

[11] Ma, S-C., and Lin, M-C., "A Trellis Coded Modulation Scheme Constructed from Block Coded Modulation with Interblock Memory", *IEEE Trans. Inform. Theory,* Vol. 40, No. 5, pp. 1348-1363, September 1994

[12] Honary, B., Kaya, L., Markarian, G.S., Darnell,M., "Maximum-Likelihood Decoding of Array Codes with Trellis Structure", *IEE Proceedings-I*, Vol. 140, No. 5, pp. 340-345, October 1993

[13] Honary, B., and Markarian, G., "New Simple Encoder and Trellis Decoder for Golay Codes", *Electronics Letters*, Vol. 29, No. 25, pp. 2170-2171, December 1993

Tables

GF(8) Symbol	Binary Representation $\alpha^2\ \alpha\ 1$	SPC Code Expansion	2*4PSK Signal Point
0	**0 0 0**	**0 0 0 0**	s_0
1	**0 0 1**	**0 0 1 1**	s_2
α	**0 1 0**	**0 1 0 1**	s_5
α^2	**1 0 0**	**1 0 0 1**	s_{13}
α^3	**0 1 1**	**0 1 1 0**	s_7
α^4	**1 1 0**	**1 1 0 0**	s_8
α^5	**1 1 1**	**1 1 1 1**	s_{10}
α^6	**1 0 1**	**1 0 1 0**	s_{15}

Table 1. GF(8) → 2*4PSK SPC-Mapping

RS Code	Rate (IBPCS)	MSED	ACG (dB)
(31,29)	**2.34**	**3.52**	**3.14**
(31,27)	**2.18**	**5.86**	**5.04**
(31,25)	**2.02**	**8.20**	**6.17**

Table 2. Bandwidth Efficient RS-Coded modulation over GF(32) with 2*8PSK modulation.

RS Code	Rate (bits/T)	Gain (dB)
(127,125)	3.44	4.81
(127,119)	3.28	9.37
(127,113)	3.11	11.36
(127,107)	2.95	12.59
(127,101)	2.78	13.43
(127,95)	2.62	14.03
(127,89)	2.45	14.48
(127,85)	2.34	14.70
(127,79)	2.18	14.95
(127,73)	2.01	15.11

Table 3. Bandwidth Efficient RS-Coded modulation over GF(128) with 2*2FSK/8PSK modulation.

RS Code	Rate (IBPCS)	MSED	ACG (dB)
(31,29)	**2.34**	**12**	**8.46**
(31,27)	**2.18**	**20**	**10.37**
(31,25)	**2.02**	**28**	**11.5**

Table 4. Bandwidth Efficient RS-Coded modulation over GF(32) with 2*8PSK modulation.

Diagrams

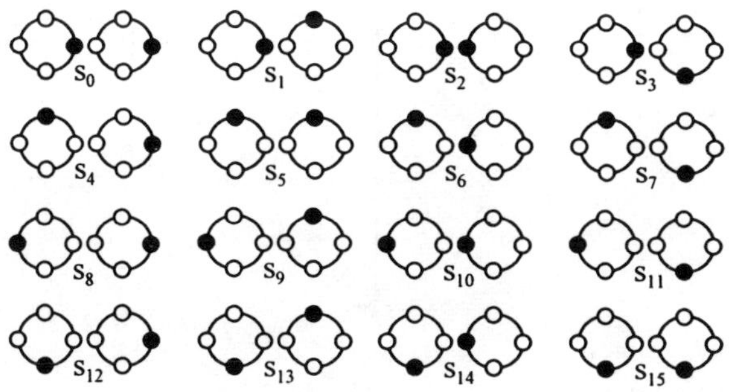

Fig. 1 2*4PSK Signal Set

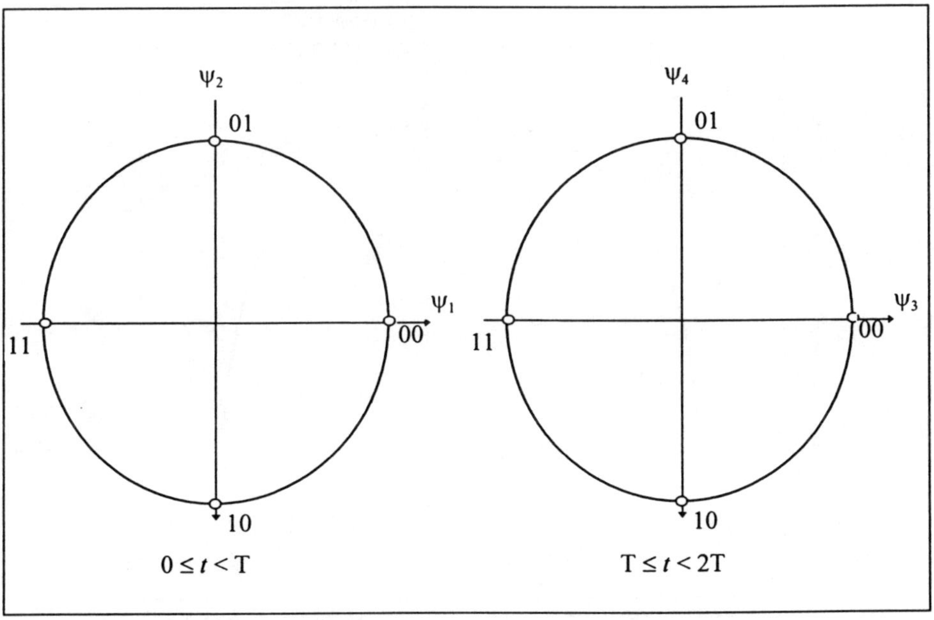

Fig. 2 Gray Code Labelled 2*4PSK Signal Constellation

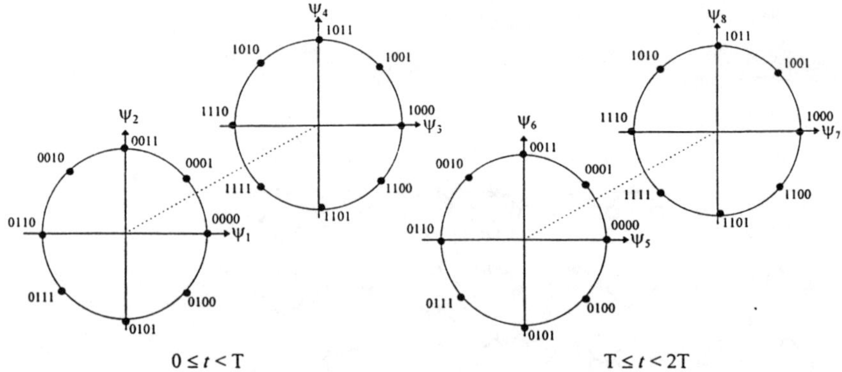

Fig. 3 2*2FSK/8PSK Signal Set

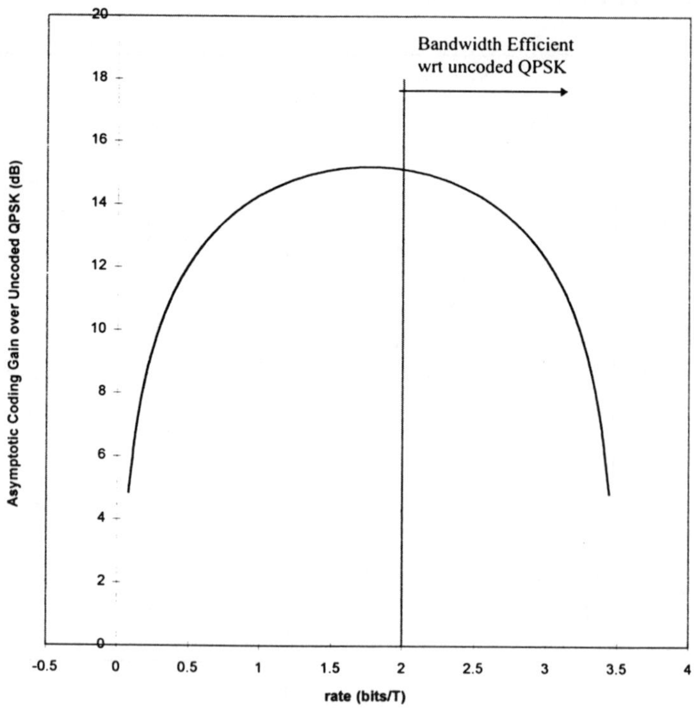

Fig. 4 Predicted Performance of Coded Modulation Schemes over GF(128)

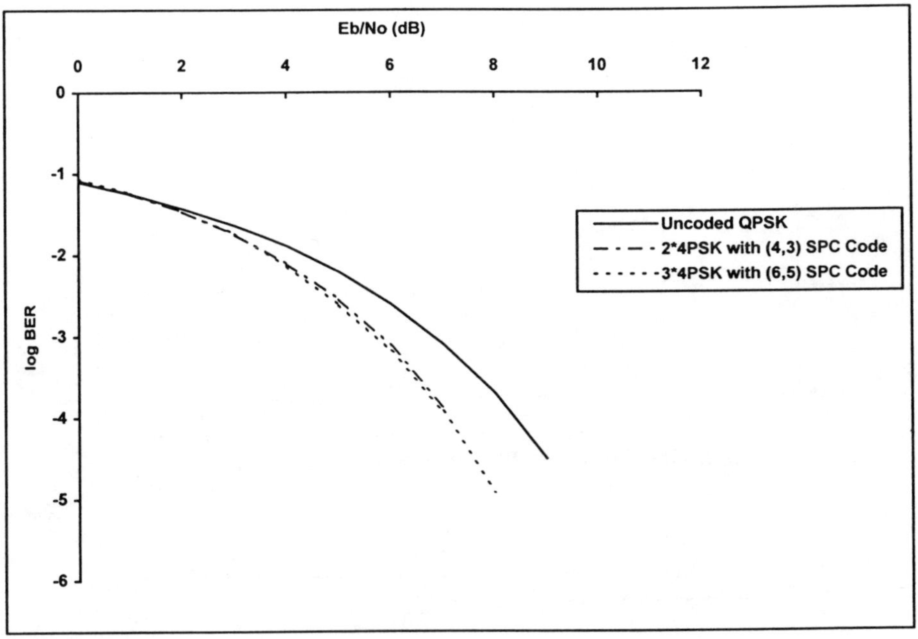

Fig. 5 N*4PSK SPC-Partitioned Modulation Scheme Simulation Results

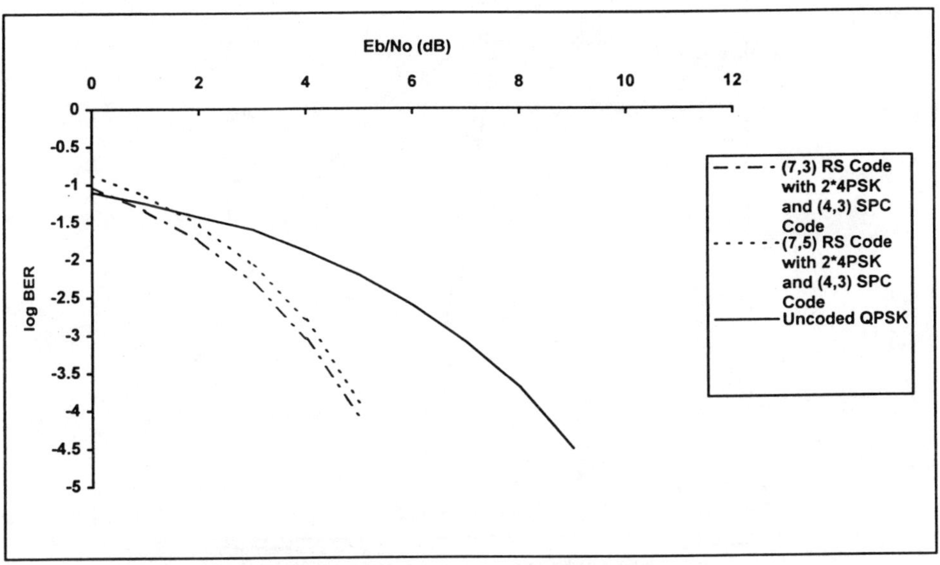

Fig. 6 RS Codes over GF(8) with 2*4PSK/(4,3)SPC Modulation

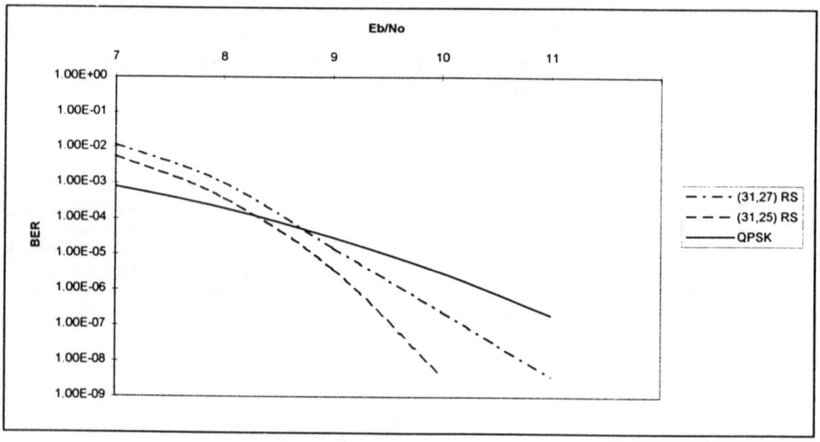

Fig. 7 RS Codes over GF(32) with 2*8PSK/(8,7) SPC Modulation

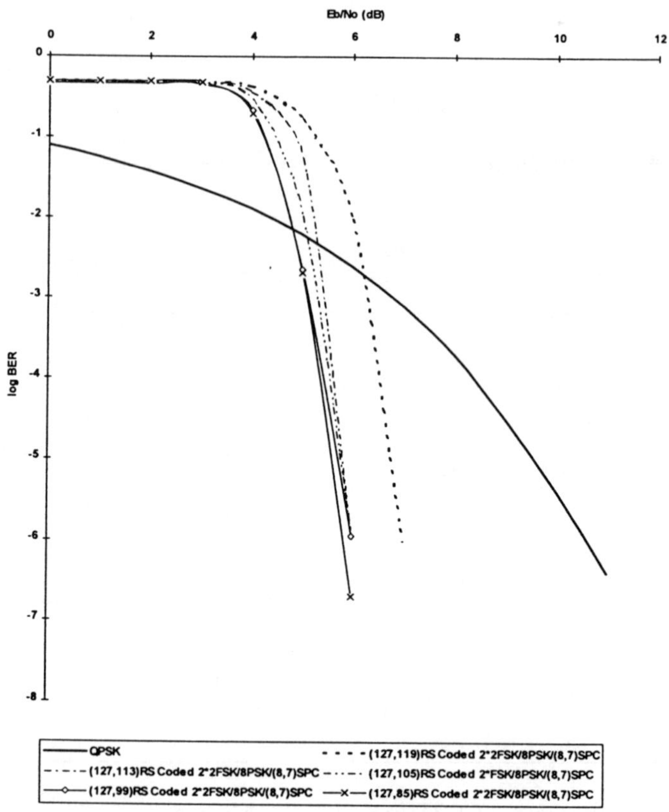

Fig. 8 Selected simulation results for bandwidth efficient RS Coded Modulation schemes over GF(128) over an AWGN Channel.

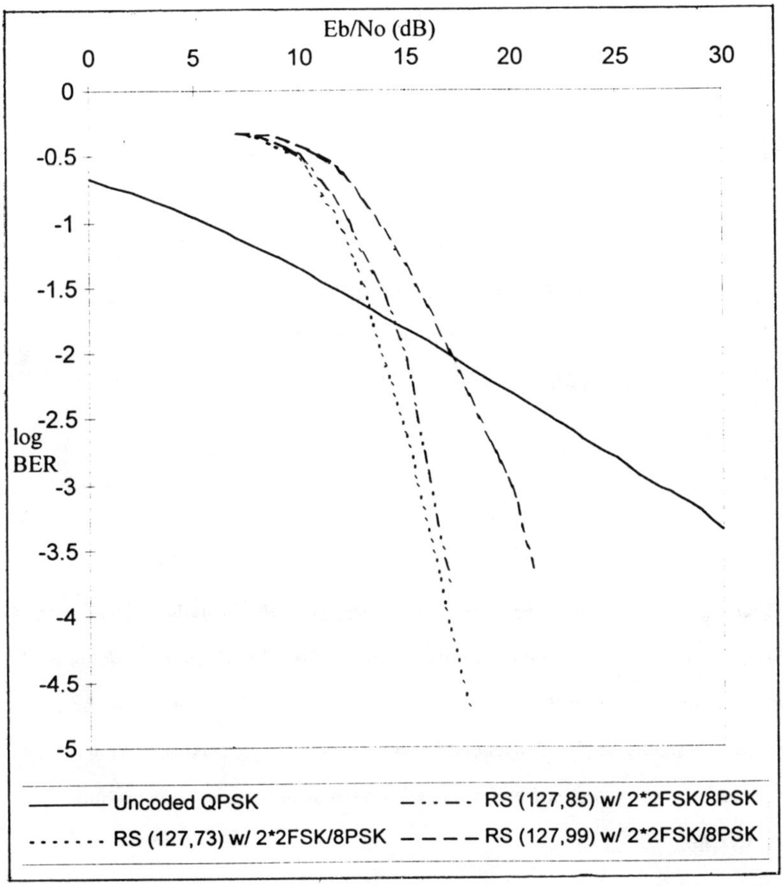

Fig. 9 Selected simulation results illustrating the performance of RS Coded Modulation over GF(128) over a Rayleigh Fading Channel

5. MTCM over Rings

Renato Baldini F.

DECOM–FEEC–UNICAMP

P. O. Box 6101

13.083-970 Campinas - SP - Brazil

Abstract

Multilevel convolutional codes defined over a ring of integers suitable for multiple trellis coded modulation (MTCM) are presented. Ring convolutional codes for MTCM can provide an extra increase in the minimum free Euclidean distance over a Ring-TCM scheme with the same number of states. As consequence, better asymptotic coding gain is reached with Ring MTCM schemes. These multilevel convolutional codes can also been seen as unity memory multilevel codes applied to MTCM.

1. Introduction

An increase in the minimum free Euclidean distance d_{free} of the trellis code relative to the minimum Euclidean distance d_{min} of a reference uncoded modulation provides better asymptotic performance gain of a trellis coded modulation system. Multiple trellis coded modulation (MTCM) can provide an extra increase in asymptotic performance gain. MTCM was first introduced by Divsalar and Simon [1] [2] and may be seeing as a coding process onto a multidimensional signal set constructed from successive channel symbols.

This work presents a MTCM scheme based on convolutional codes defined over a ring of integers modulo-M, where M is a integer. These codes are similar to unity memory codes and can provide an extra increase in the minimum free Euclidean distance over the equivalent uncoded modulation scheme.

Before introducing the Ring MTCM, a brief introduction on multiple TCM is presented in the next section. In addition, a simple example of MTCM is also given.

2. Multiple TCM [3]

Multiple Trellis Coded Modulation (MTCM) [1] can be implemented by an encoder with b input bits and s output binary sequences which are mapped into k M-ary symbols in each transmission interval. The parameter k is known as the *multiplicity* of the code and it represents the number of M-ary symbols allocated to each branch of a trellis diagram. Fig. 1 shows a block diagram for a generalized MTCM encoder.

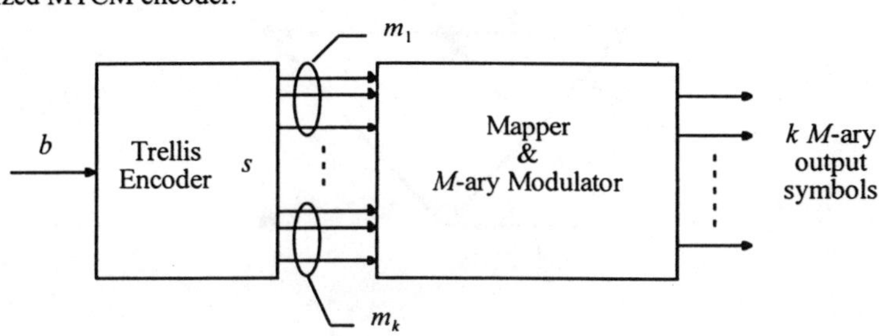

Fig. 1 Generalized MTCM Encoder

The s trellis encoder output bits are grouped into k groups, each of them containing m_1, m_2, …, m_k bits. Each group of bits is then mapped into an M_i-ary output symbol, where $M_i = 2^{m_i}$, $i = 1, 2, \ldots, k$. The throughput r is defined as the ratio of the bit rate to transmission bandwidth, that is, $r = b/k$ bits/s/Hz.

Divsalar and Simon [1] [2] [3] have proposed a MTCM technique for the case where $b = mk$ and $s = (m + 1)k$ with k groups of $(m + 1)$ bits each being mapped into an M-ary symbol chosen from an alphabet of size $M = 2^{m+1}$. In an transmission interval, mk bits enter the trellis encoder and k M-ary symbols leave the mapper. The throughput is therefore m bits/s/Hz that results in an unity-bandwidth expansion relative to an uncoded modulation with 2^m signal points.

Example 1: Two-State MTCM [3]

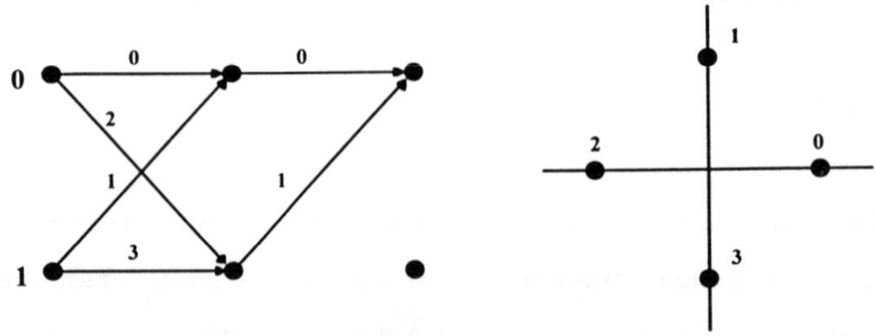

a) Rate 1/2 Trellis Coded 4-PSK

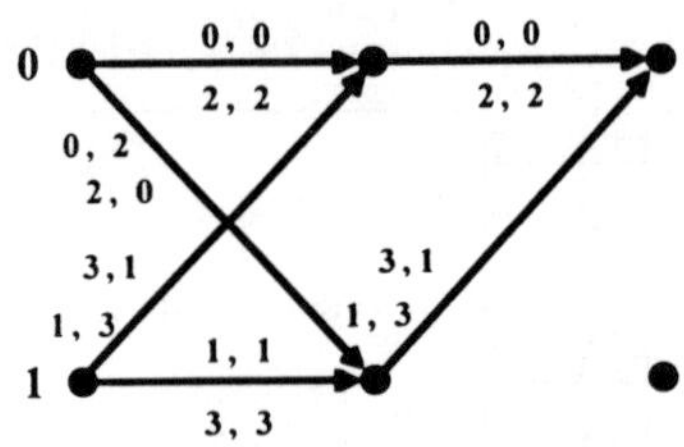

b) Rate 2/4 Multiple Trellis Coded 4-PSK, $k = 2$

Fig. 2 Trellis Diagram for Conventional and Multiple TC-4-PSK Modulation

Fig. 2(a) shows a trellis diagram for the conventional rate 1/2 trellis coded 4-PSK. If the average energy of the constellation 4-PSK is made equal the unity, the achievable minimum free Euclidean distance of this scheme is equal to 6.

Fig 2(b) is a trellis diagram for the corresponding multiple trellis coded 4-PSK, where $b = 2$, $m = 1$ and $k = 2$. There are $2^{mk} = 4$ transition from each of the two states, therefore there must be two parallel transitions between two states. Two output 4-PSK symbols are assigned to each transition. The squared free Euclidean distance achieved with this 2-state MTCM is 8.

3. MTCM over Rings

Fig. 3 shows a generalized Ring-MTCM encoder. The Ring-MTCM is implemented by a Gray mapper with b input bits and jk output M-ary symbols belonging to a ring of integers modulo-M ($j > 1$). The jk M-ary symbols are then encoded into jn M-ary symbols also belonging to the same ring of integers modulo-M. In analogy to MTCM, the parameter j can be defined as the *multiplicity* of the code and jn represents the number of M-ary symbols assigned to each branch of a trellis diagram. The number b of input bits depends on the number M of the PSK modulation signal symbols and the number jk of input symbols of the multilevel trellis encoder.

Conventional convolutional codes over rings applied to coded M-PSK modulation are generally (n, k, m) multilevel codes where n is made equal to $k+1$ in order to compare with an equivalent uncoded modulation scheme and m is the number of memories of the code.

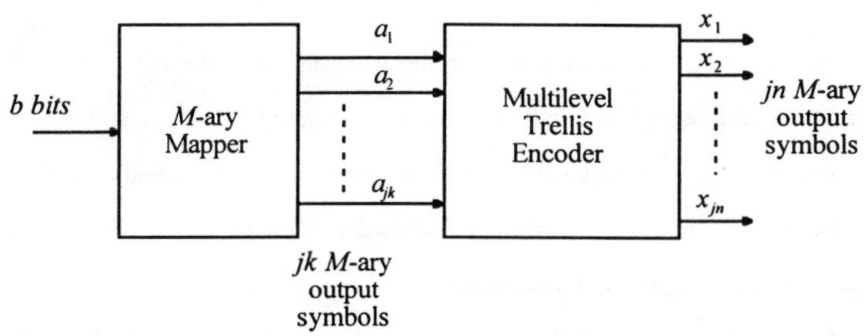

Fig. 3 Generalized Ring MTCM Encoder

Example 2: MTCM based on a Ring of Integers Modulo-4

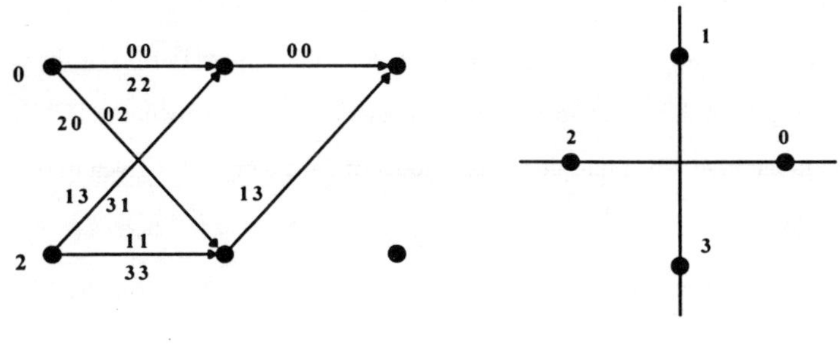

a) Two-State Rate 1/2 Trellis Coded 4-PSK over a Ring of Integers Modulo-4 (Conventional)

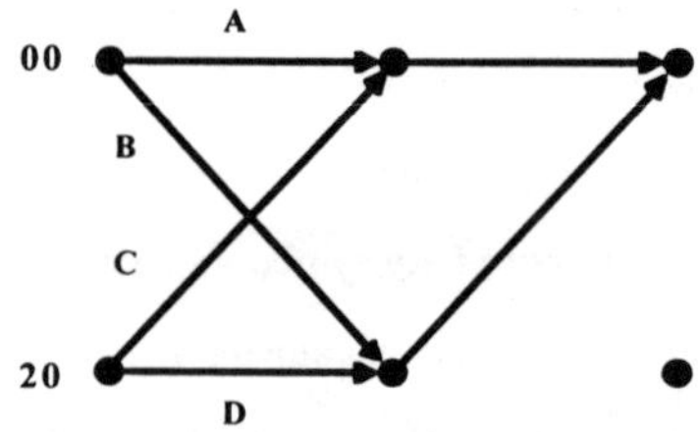

A	B	C	D
0000	0131	0020	0111
0222	0313	0202	0333
1032	1123	1012	1103
1210	1301	1230	1321
2020	2111	2000	2111
2202	2333	2222	2313
3012	3103	3032	3123
3231	3321	3211	3301

b) Two-State Rate 2/4 Multiple Trellis Coded 4-PSK over a Ring of Integers Modulo-4

Fig. 4 Trellis Diagram of Conventional and Multiple TC-4-PSK over a Ring of Integers Modulo-4

Fig. 4(a) shows a trellis diagram for the conventional rate 1/2 trellis coded 4-PSK defined over a ring of integers modulo-4. The average energy of the constellation 4-PSK is made equal the unity therefore the achievable minimum free Euclidean distance of this scheme is equal to 8. Notice that this coded modulation scheme has the same Euclidean distance of the two-state rate 2/4 MTCM given in the example 1.

Fig. 4(b) is a trellis diagram for the corresponding multiple trellis coded 4-PSK defined over a ring of integers modulo-4, where $b = 4$, $j = 2$, $k = 1$ and $n = 2$. There are $4^{jk} = 16$ transition from each of the two states, therefore there must be 8 parallel transitions between two states. Four output 4-PSK symbols are assigned to each transition. The squared free Euclidean distance achieved with this 2-state Ring MTCM is also equal to 8. Therefore, for this simple example there is no increase in the Euclidean distance of the Ring MTCM over the conventional Ring TCM. Despite of this no increase in performance, the two state Ring MTCM given in the example 2 is illustrative. The multiple trellis encoder of the example 2 is given in fig. 5.

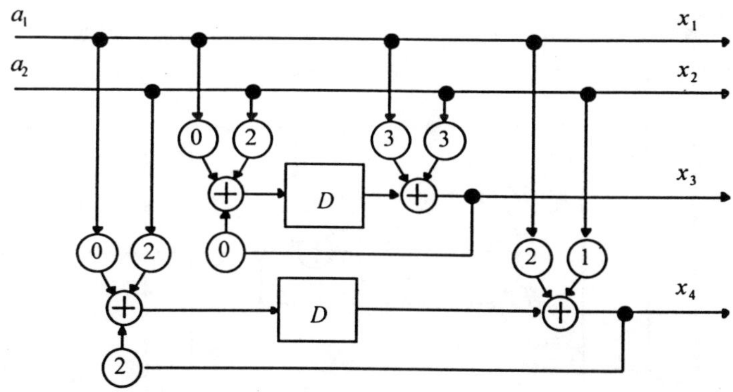

Fig. 5 Two-State Multiple Trellis Encoder Structure defined over a Ring of Integers Modulo-4

Notice that the second memory in diagram given in fig. 5 is always at the state equal to 0, whereas the first memory can reach only the two states 0 and 2. Therefore the trellis diagram can reach only two states, that is, the state 00 and 20.

Fig. 6 shows the general structure of a multilevel convolutional encoder for convolutional codes defined over rings of integers modulo-*M*. Notice that the multilevel convolutional codes are all made systematic in order to avoid catastrophicity. Notice that the content of a given memory is independent of the contents of the other memories, i. e., there is no interconnection among memories. The encoding scheme shown in Fig. 6 is very similar to the well-known unity memory encoders.

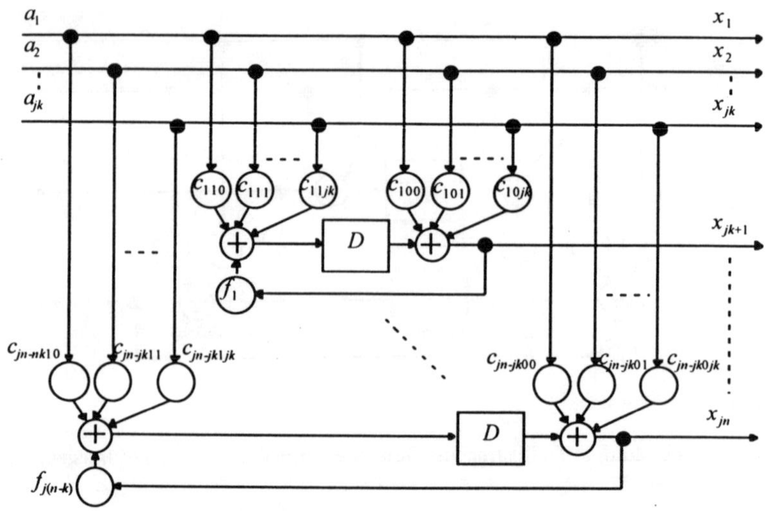

Fig. 6 General Structure of a Multilevel Convolutional Encoder

4. Results

Table 1 shows some multilevel convolutional codes with rate $j/2j$ ($j = 2, 3$) defined over a ring of integers modulo-4 that present better minimum free Euclidean distance than an equivalent rate 1/2 Ring-TCM codes. The constellation utilized in those schemes is the 4-PSK modulation.

Table 1: Minimum Euclidean Free distance of Some Rate $j/j2$ MTCM defined over a Ring of Integers Modulo-4

Rate	States	MTCM over Z_4	Conventional Rate 1/2 TCM over Z_4
2/4	2	8	8
2/4	4	10	10
2/4	8	16	12
2/4	16	16	16
3/6	32	18	16
3/6	64	20	18

Notice from the table 1 that for number of states greater than 8, the MTCM over Z_4 presents a gain in terms of minimum free Euclidean distance over the conventional rate 1/2 TCM defined over Z_4. The Ring TCM schemes has similar performance of the Ungerboeck TCM, therefore, one may conclude that the Ring MTCM schemes presented in table 1 has also better performance over the Ungerboeck TCM. The generalization for higher order PSK modulation is straightforward.

5. Conclusion

MTCM defined over rings is easier to be searched than those presented by Divsalar and Simon [1] [3]. Moreover they can be easily made transparent to phase rotation of the carrier by searching for convolutional codes that contain the all one sequence as a codeword.

It is also possible to find convolutional codes defined over a ring of integers modulo-*M* for MTCM that utilizes PSK modulation with higher number of signal points than those presented in this work. By using a simple adaptation it is possible to use QAM modulations in the Ring MTCM schemes.

Acknowledgments

I would like to thank Dr. Jaime Portugheis for some very useful discussions during the realization of this work.

References

1. Divsalar, D. and Simon, M. K., "Multiple trellis-coded modulation (MTCM)", IEEE Trans. Comms., Vol. 36, No 4, pp. 410-419, April 1988.

2. Divsalar, D. and Simon, M. K., "Generalized multiple trellis-coded modulation", ICC'87 Conf. Rec., Seattle, Wash., pp. 20.3.1 - 20.3.7, June 7-10, 1987.

3. Biglieri, E. Divsalar, D., McLane, P. J. and Simon, M. K., *An introduction to Trellis-Coded Modulation with Applications*, Chapter 7, Macmillan, 1991.

4. Baldini F., R. and Farrell, P. G. , "Coded modulation based on rings of integers modulo-*q*, Part-2: Convolutional codes", IEE Proceedings Comms., Vol. 141, No 3, pp. 137-142, June 1994.

CHAPTER 2

Encryption and Sequence Design

6. Cryptography: an Introductory Overview

Fred Piper

Acknowledgment

An introductory overview of cryptography might seem 'out of place' at a specialist coding theory conference. However, the main objective of this particular conference is to acknowledge the work of Paddy Farrell and his contribution to the early development of academic cryptography in the UK deserves recognition. He was one of the prime movers in the establishment of the series of IMA conferences on Coding and Cryptography which were introduced to assist research students in the two disciplines and to encourage interaction between them. Most of the early cryptographers in the UK were mathematicians and we are indebted to Paddy for his engineering input to their development.

Encryption algorithms

The basic aim of an encryption algorithm is to 'scramble' messages so as to make them unintelligible to eavesdroppers. In some sense the encryption process can be regarded as adding noise to the data with the declared intention that the attacker should not be able to correct the errors. Ideally the probability of any bit being corrupted is 0.5 and this is not the type of noisy channels that coding theorists usually confront. The genuine receiver then uses the appropriate secret key decryption algorithm to reverse the process and obtain the clear message. This is shown diagrammatically in Figure 1.

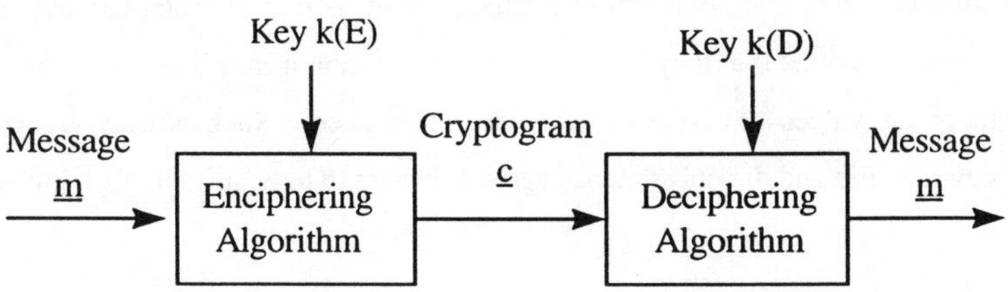

A Cipher System
Figure 1

In most practical scenarios the algorithm will be known to attackers. However, the deciphering key is kept secret and it is knowledge of this key which gives the receiver an 'advantage' over the attackers. Unless it provides information about k(D), knowing k(E) does not really assist the attacker and this is one of the reasons why encryption algorithms are divided into two disjoint categories; symmetric and asymmetric.

An encryption algorithm is **symmetric** if knowledge of k(E) makes it easy to determine k(D). In fact in many symmetric systems they are identical. If, however, it is computationally infeasible to determine k(D) from k(E) then the system is **asymmetric**. In this latter situation, since it does not facilitate an attack on the algorithm, k(E) can be (and frequently is) made public. In fact such algorithms are often known as **public key** algorithms.

The use of a symmetric algorithm requires some form of business relation or trust between the two communicators since they need to share some secret information. In essence the users are relying on the cryptography to protect them from all third parties. However, there is no requirement for protection from each other. If they do not trust each other then there is a need for some form of asymmetry between them and they may be forced to use a public key system. Even if they do trust each other, the establishment of these secret keys before the transmission of the secret message can be a nightmare. For this, and many other reasons, key management is probably the most difficult aspect of establishing a secure communication link. One of the main attractions for using an asymmetric system is that there is may be no need to distribute any secret keys. However, it is important to note that public keys do need 'protection'. If, for instance, I were able to convince someone that a public key for which I knew the secret key belonged to everyone else then I would be able to read all encrypted messages sent to that person. Public keys need authentication i.e. it is important that everyone has confidence in the identity of their owners.

If the algorithm is known then one option for the attacker is that of systematically trying all deciphering keys. Provided that they have suitable information it may be possible to eliminate obviously incorrect values until the correct one is obtained. Such attacks are known as exhaustive key searches and the only defence against them is to have sufficiently many keys that they require too much resource and/or time. Trying to quantify the number of keys required is, of course, not an exact science as it depends upon a number of parameters including, for instance, the value of the data being protected.

When choosing an encryption algorithm to provide secrecy, one fundamental decision is whether to encrypt the data bit by bit or to divide it into blocks and encrypt by blocks. Ciphers which encrypt bit by bit are known as stream ciphers. In this situation the decision about whether or not a bit is changed must depend upon its position in the bit stream. Thus the encryption process involves XORing a keystream sequence with the data (See Figure 2).

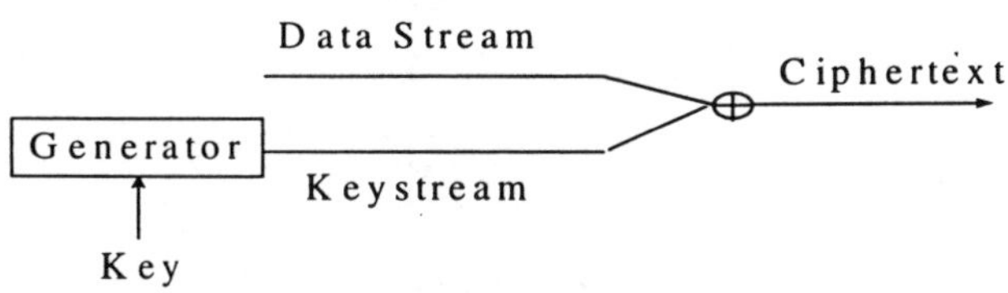

A Stream Cipher
Figure 2

Encryption algorithms for stream ciphers are essentially nothing more than pseudo-random binary sequence generators and, as such, have many other applications including, for instance, frequency hoppers. The problems associated with designing 'good' generators include determining the properties that they should have, deciding how to test for those properties and then designing efficient algorithms to achieve these objectives. There are many definitions of pseudo-randomness and a number of statistical tests have been devised to appraise the suitability of generators. Since any finite bit string can be generated on a linear feedback shift register the linear complexity of a sequence, i.e. the length of the shortest LFSR which can generate it, has particular cryptographic significance and the celebrated Berlekamp-Massey algorithm, which started life as a decoding algorithm for BCH codes, is a fundamental part of the stream cipher designer's repertoire.

One major attraction of stream ciphers is that the encryption process does not lead to any error propagation. This is extremely relevant for communications over noisy channels and devises for the encryption of digitalised speech, as in GSM mobile telephone networks, tend to use them. They are also very fast. One major drawback is that, since the encryption and decryption processes both need the same sequence, it is necessary to ensure that the two generators are always in step. This necessitates the overhead of transmitting synchronisation data for no reason other than re-initialising the two generators.

For a block cipher the diagram is somewhat different as is shown in Figure 3.

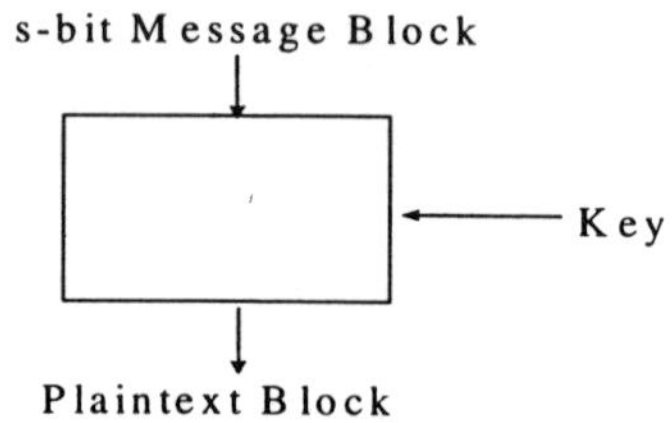

A Block Cipher
Figure 3

If a message is encrypted by dividing it into blocks and, after choosing a key, putting it through the encryption algorithm then identical message blocks will result in identical ciphertext blocks. The method of using an encryption algorithm in this way is known as Electronic Code Book (ECB) mode and is very rarely used for long messages because it is prone to statistical attacks and does nothing to prevent an attacker manipulating the data by, for instance, changing the order of the ciphertext blocks, deleting or repeating specific ciphertext blocks etc. To prevent these attacks it is usual to use the algorithm in such a way that the ciphertext block of any given plaintext block depends not only on the contents of the block but also on its position in the message or even the contents of all the previous blocks. An example of one such mode of operation is shown in Figure 4. This particular mode of operation is known as 8-bit Cipher Feed Back (CFB) and it is interesting to note that when the ciphertext is decrypted the algorithm is used in encrypt rather than decrypt mode and that the enciphering and deciphering keys are identical.

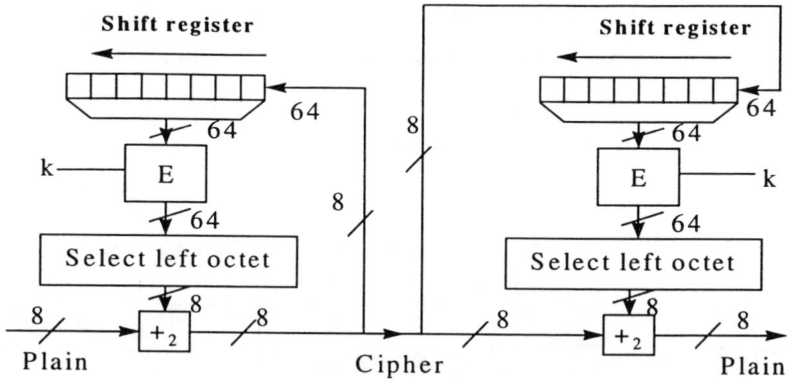

Cipher Feedback Mode
Figure 4

Whenever a block cipher is used then it is an important aspect in the design that any (small) change in either the key or the data should result in an unpredictable change in the ciphertext block. Thus error propagation is almost a design criterion and block ciphers should not be used is situations where error propagation is undesirable.

Note that if the algorithm is used in CFB mode as illustrated in Figure 4 then, although each ciphertext bit depends on every bit of the preceding message, if one bit of the ciphertext is in error the deciphering process is self-synchronising in the sense that, after one wrong bit and an unreliable 64-bit block the plaintext will be correct.

So far we have assumed that encryption is being used to provide privacy. While this is the natural way to introduce the topic, the provision of privacy is not the main commercial use of cryptography. Instead it is used to provide data integrity and user authentication. One of the main applications of block ciphers is the provision of cryptographic checksums to detect deliberate manipulation of the message. A typical instance would be where the message was divided into blocks. The first block would be encrypted and each subsequent block would be XORed with the previous cipher block before being encrypted itself. All ciphertext blocks except the last are discarded and the message is sent in clear with part of the final ciphertext block appended as the checksum. In many banking systems DES, a block cipher with 64-bit blocks is used but the checksum, which is known as a MAC, is only 32 bits. This is a clear illustration of how important error propagation can be.

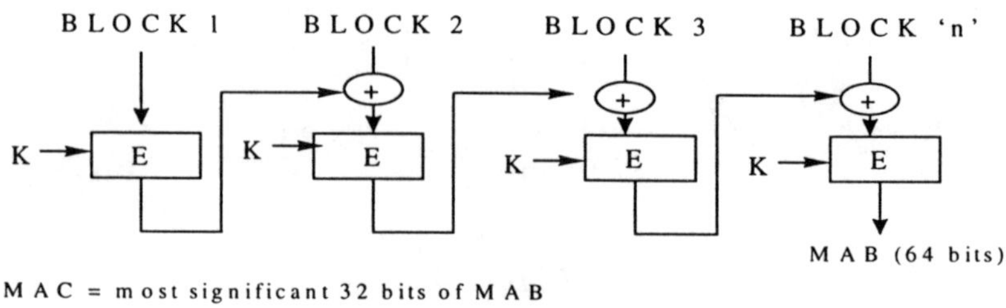

MAC Generation
Figure 5

If a symmetric algorithm is used to provide a cryptographic checksum then either the sender or recipient can alter the message and adjust the checksum so that it is correct for the altered message. Whenever there is a need to settle disputes between sender and receiver or, in particular, a requirement that the sender should not be able to deny that they sent a particular message then some form of asymmetry between sender and receiver is required. Indeed the cryptographic checksum should be such that only the sender could have computed it. If this is to be guaranteed by the cryptography then the checksum must depend on a secret value known **only** to the sender. Given this requirement it is natural to try to use the secret deciphering key of an asymmetric algorithm to compute the checksum which, in this case, is called a **digital signature**. If this is done then anyone who knows the corresponding public encryption key can use it to verify the signature. (Note: it is not possible to use all public key systems to provide signatures in precisely this way. However, this is how RSA digital signatures 'work'.)

If I wish to forge someone else's digital signature then either I must obtain their secret key or I must convince other people that my public verification key really belongs to that person. If I can do this then I can use my signing key and all third parties will accept the signature as genuine. The use of a sufficiently strong cryptographic algorithm will prevent attackers computing secret signing key values from public verification keys. Unfortunately the problem of ensuring the authenticity of users' public keys is not so straightforward. Current practice is to establish a Public Key Infrastructure in which Certification Authorities (CAs) are established . These CAs issue (digitally) signed certificates to verify that a public key value belongs to a specific user. If this is done then anyone who has an authentic copy of the CA's public key can check the signature on the certificate and thereby have confidence in the identity of the owner of the public key value. Clearly this process is totally reliant on the procedures which the CA has in place to verify the identity of the user during the certification process. It also relies on the existence of a mechanism for enabling users to have confidence that their copy of the CA's public key is genuine.

Whilst there are numerous interesting technical cryptographic problems to be solved, there is little doubt that one of the main barriers to secure electronic commerce is the lack of suitable infrastructures to allow reliable user authentication and key management.

Royal Holloway
Mathematics Department
Egham, Surrey TW20 0EX

Tel: 01784 443101
Fax: 01784 430766
Email: f.piper@rhbnc.ac.uk

7. Two-Dimensional Interleaving

Valdemar C. da Rocha Jr.
Communications Research Group - CODEC
Department of Electronics and Systems
Federal University of Pernambuco
50741-540 Recife PE BRASIL

Abstract

Interleaving is a venerable technique used in conjunction with error-correcting codes to combat errors in channels with memory. Information digits appearing in two-dimensional arrays, e.g., magnetic tapes, discs, chips, etc. need to be protected against such errors, which are called patch or cluster errors. The concept of a compact patch of errors is introduced and the correction of multiple compact patches of errors in two-dimensional arrays is described by using random error-correcting codes in one array dimension combined with permutation of coordinates in the other array dimension.

1 Introduction

This paper is dedicated to Patrick G. Farrell, on the occasion of his sixtieth birthday, in recognition of his own long interest in practical applications of coding theory and in particular in the construction and practical application of array codes [1], and to whom we owe our introduction to the field of error-correcting codes.

Many authors have studied the problem of correcting two-dimensional bursts of errors (also called clusters or patches of errors) in the case where the patch of errors has either a square or rectangular shape. The digit arrays considered are in general rectangular $m \times n$, where both m and n are positive integers. We refer the reader to [1] for a thorough reference list on this problem. More recently Almeida and Palazzo [2] dealt with the correction of patches of errors of symmetric shape, not necessarily rectangular, in square arrays, i.e., arrays for which $m = n$, by employing a linear transformation of the array coordinates to achieve the desired interleaving. Their treatment was however mostly heuristic and the code parameters were determined by computer search. In a subsequent paper Kauffman et alli [3] proved part of the results in [2] also for the case where the array was a square with side $n = rs$, where both r and s are positive integers, and where the patch of errors had a

*This work received partial support from the Brazilian National Council for Scientific and Technological Development (CNPq) under the grant No. 304214/77-9.

rectangular $r \times s$, $r \neq s$ shape, or a square shape if $r = s$. In this paper we will consider rectangular $m \times n$ arrays of digits, where both m and n are positive integers. We remark that the current trend to deal with this problem is to employ coding along the two dimensions of the array, combined with an interleaving transformation which changes both coordinates of most of the array digits. Contrary to this trend, we will employ coding along a single array dimension, combined with an interleaving transformation which acts only in the other array dimension. This approach not only turns out to be simpler but also paves the way for establishing a simple solution to the problem of correcting multiple patches of errors.

2 Definitions and Preliminaries

Here and hereafter whenever we refer to an array we will mean a two-dimensional array. Let us denote by (n, k) a linear block code of blocklength n and dimension k. We will consider an $m \times n$ array whose m rows are codewords of an (n, k) linear error-correcting block code. Let us denote by (x_i, y_i), $0 \leq x_i \leq n-1$, $0 \leq y_i \leq m-1$, the coordinates of the array digits, i.e., the row positions vary from 0 to $n-1$ while the column positions vary from 0 to $m-1$. The *encoded* array is then interleaved, for either storage or transmission purposes, by permuting the digits in each column as follows.

Definition 1 *We define the matrix* T

$$T = \begin{bmatrix} 1 & -b \\ 0 & b \end{bmatrix},$$

where b *is a positive integer relatively prime to* m*, and perform the interleaving of the digit positions* (x_i, y_i)*,* $0 \leq x_i \leq n-1$*,* $0 \leq y_i \leq m-1$*, in a rectangular* $m \times n$ *array* A*, by means of the linear transformation*

$$(x_i, y_i) \begin{bmatrix} 1 & -b \\ 0 & b \end{bmatrix} = (x_i, -bx_i + by_i) = (x'_i, y'_i),$$

where (x'_i, y'_i)*,* $0 \leq x'_i \leq n-1$*,* $0 \leq y'_i \leq m-1$*, denotes the digit positions in the resulting interleaved array* A'*.*

Obviously, the above transformation preserves the first coordinate, i.e., $x'_i = x_i$, and acts only on the second array coordinate. This is equivalent, for a fixed value of the first coordinate, to perform permutations of the elements in the second coordinate in the $m \times n$ array A in order to obtain A'. We will refer to a transformation like the one above, which acts on a single coordinate, as *simple interleaving.*

We felt that a precise definition of a patch of errors was needed before we could start speaking of patch error correction. Although there are many possible definitions for a patch of errors, as there are for bursts of errors in serial data transmission, we found it reasonable to consider a definition that takes into account how close the error digits are in a patch of errors rather than to take into account a specific shape for the patch. We will therefore consider in our definition of a patch of errors the Lee distance d_L between pairs of array positions. The

Lee distance [4, p.41] between any two array positions with coordinates (x_1, y_1) and (x_2, y_2), respectively, is defined as

$$d_L = \min\{|x_2 - x_1|, |m - (x_2 - x_1)|\} + \min\{|y_2 - y_1|, |n - (y_2 - y_1)|\}.$$

Definition 2 *We define a d_L-patch of errors in an $m \times n$ array to be any set of error digits whose pairwise Lee distance is at most d_L.*

Definition 3 *We define the weight of a patch of errors as the number of erroneous digits in the patch.*

Definition 4 *We define a matched patch of errors in an $m \times n$ array to be the weight m d_L-patch of errors of minimum d_L for the given m.*

We say that a patch of errors P_2 is *covered* by a patch of errors P_1 if any error position in P_2 is contained in P_1.

Definition 5 *We define a compact patch of errors in an $m \times n$ array to be any two-dimensional error pattern of weight at most m which is covered by a matched patch of errors.*

Compact patches of errors play a key role in our work because, as we will show later, they constitute the patches of errors guaranteed to be correctable in our proposed *simple interleaving* scheme. Of course patches of errors, other than compact patches, may also be correctable by our scheme but that comes as an extra bonus rather than as a result of design.

Lemma 1 *The de-interleaving transformation T^{-1}, obtained from T, is given by*

$$T^{-1} = \begin{bmatrix} 1 & 1 \\ 0 & b^{-1} \end{bmatrix},$$

where b^{-1} is the inverse of b modulo m.

Proof: We assumed that b is relatively prime to m and therefore b^{-1} exists. It is now immediate to check that $TT^{-1} = I_2$, where I_2 is the unit matrix of order 2. □

3 Spreading and Correction

Perfect interleaving of a weight n compact patch of errors in a square $(n \times n)$ array has been defined in [2] to be that interleaving for which no two or more errors in the patch occupy the same row or the same column in the square array after being interleaved. Applying the above definition to an $m \times n$ array we observe that perfect interleaving of a given patch of errors can only be achieved if its maximum weight does not exceed $\min\{m, n\}$. We noticed however that, if we are using a t-error-correcting code to encode the array rows (before interleaving) we do not need to worry about perfect interleaving. All we need to guarantee is that after de-interleaving each array row contains at most t positions in error, i.e., if we decode the array row-by-row the number of errors in any given column will not affect our decoding procedure. However, if we were using codes in more than one array dimension as, for example, row and diagonal single-parity-check codes, then it would be essential to employ perfect interleaving. For our purposes we change slightly the definition of perfect interleaving as follows.

Definition 6 *We define perfect spreading of a patch of errors of weight at most m, associated with an $(m \times n)$ array A' obtained by simple interleaving, to be that de-interleaving for which no two or more errors in the patch occupy the same row in the array resulting after de-interleaving.*

We next establish a relationship between the integer b, used in the transformation matrix T, and the error digit positions in a patch of errors of weight at most m, associated with a given $m \times n$ array, concerning the existence of a perfect spreading.

Theorem 1 *Let A' denote an $m \times n$ array interleaved by means of T, possibly corrupted by patches of errors. Let T^{-1} denote the corresponding de-interleaving transformation. A perfect spreading exists if and only if there is an integer b such that*

$$(x_2 - x_1) + b^{-1}(y_2 - y_1) \not\equiv 0 \text{ mod } m \tag{1}$$

for any pair of digit positions (x_1, y_1) and (x_2, y_2) belonging to a patch of errors of weight at most m, associated with the $m \times n$ array.

Proof: Suppose on the contrary that (1) holds but that the spreading is not perfect. Denoting by (x_1, y_1) and (x_2, y_2) any two positions in a patch of errors of weight at most m and applying the transformation T^{-1} to each one of them we obtain

$$(x_1, y_1)\begin{bmatrix} 1 & 1 \\ 0 & b^{-1} \end{bmatrix} = (x_1, x_1 + b^{-1}y_1) \tag{2}$$

$$(x_2, y_2)\begin{bmatrix} 1 & 1 \\ 0 & b^{-1} \end{bmatrix} = (x_2, x_2 + b^{-1}y_2). \tag{3}$$

If the transformed positions $(x_1, x_1 + b^{-1}y_1)$ and $(x_2, x_2 + b^{-1}y_2)$ occupy the same array row then it follows from (2) and (3) that

$$x_1 + b^{-1}y_1 \equiv x_2 + b^{-1}y_2 \text{ mod } m$$

or

$$(x_2 - x_1) + b^{-1}(y_2 - y_1) \equiv 0 \text{ mod } m,$$

which is a contradiction to inequality (1). Hence (1) is a sufficient condition to ensure perfect spreading of the digits in a patch of errors of weight at most m. On the other hand if we assume that (2) and (3) represent positions in two distinct array rows then it follows that

$x_1+b^{-1}y_1 \not\equiv x_2+b^{-1}y_2$ and (1) is verified to be also a necessary condition for perfect spreading of the error digits in a patch of errors of weight at most m, and the theorem follows. □

Given that (1) holds and since the transformation T is reversible, we conclude that if t or less patches of errors of weight at most m affect the $m \times n$ array A', after de-interleaving each row of the array A will contain at most t digits in error. We then decode the de-interleaved array using the error-correcting properties of the (n, k) code before delivering the data to a data sink. We have thus proved the following.

Theorem 2 *Let A' be the $m \times n$ array obtained from the array A by the transformation T, possibly corrupted by patches of errors. Let the rows of A be codewords of an (n, k) t-error-correcting code and suppose that b satisfies Theorem 1. By means of the de-interleaving transformation T^{-1} any pattern of at most t patches of errors of weight at most m, in A' is correctable.*

We are aware that unless we are more specific about the patches of errors meant to be correctable by *Theorem 1* it may be either hard to find a b or else, for a given b, the set of correctable patches may be too small thus making the proceduce ineffective. Therefore it seems wise to constrain the patches of errors to guarantee that inequality (1) holds for at least those patches considered typical in a given application. For this reason from now on we will consider only compact patches of errors.

Lemma 2 *The Lee distance d_L between pairs of error digit positions in a compact patch of errors satisfies the following inequality*

$$d_L \leq -1 + \sqrt{2m-1}.$$

Proof: The number N_{d_L} of positions with Lee distance at most d_L from an arbitrary but fixed array position is given by

$$N_{d_L} = 1 + 2d_L(d_L + 1).$$

It follows that a d_L-patch of errors has weight at most N_{d_L}. Furthermore, by definition, a compact patch of errors has weight at most m, which means that we must have $N_{d_L} \leq m$, or

$$2d_L^2 + 2d_L + 1 \leq m. \tag{4}$$

Solving (4) for d_L we obtain the desired result and the lemma follows. □

Theorem 3 *Let (x_1, y_1) and (x_2, y_2) denote any two digit positions in a compact patch of errors associated with an $m \times n$ array. These compact patches of errors are correctable by simple interleaving if and only if there is a positive integer b which satisfies the following condition*

$$|t|_L + |b^{-1}t|_L > -1 + \sqrt{2m-1},$$

where t is a positive integer, $1 \leq t \leq -2 + \sqrt{2m-1}$, and where $|.|_L = d_L(.,0)$ mod m.

Proof: Without loss of essential generality let us suppose that (x_1, y_1) is a solution of the congruence $x + b^{-1}y \equiv c \bmod m$. It follows that for any other solution (x, y) of this congruence we can write [5, p.5-6]

$$(x - x_1) + b^{-1}(y - y_1) \equiv 0 \bmod m,$$

or

$$\begin{aligned} y - y_1 &\equiv t \\ x - x_1 &\equiv -b^{-1}t, \end{aligned}$$

where t is an integer. Therefore the Lee distance d_L between (x, y) and (x_1, y_1) can be written as

$$d_L = |t|_L + |b^{-1}t|_L.$$

We notice however that by *Theorem 1* the error digits in a compact patch of errors must satisfy inequality (1). Thus we conclude that (x, y) and (x_1, y_1) can not both belong to the same compact patch of errors. Consequently, by *Lemma 2* their pairwise Lee distance must satisfy $d_L > -1 + \sqrt{2m-1}$, i.e., $|t|_L + |b^{-1}t|_L > -1 + \sqrt{2m-1}$. Any value of b satisfying the last inequality for a positive t, $1 \leq t \leq -2 + \sqrt{2m-1}$, simultaneously guarantees that pairs of positions satisfying the congruence do not belong both to the same compact patch of errors and that their Lee distance is greater than the Lee distance between any pair of error positions in a compact patch.

□

Example
We consider the 5×7 array depicted in the top of figure 1, whose rows we assume to be codewords of a single-error-correcting code, and the following transformation matrix T

$$T = \begin{bmatrix} 1 & 3 \\ 0 & 2 \end{bmatrix}.$$

The resulting interleaved array is shown in the bottom of figure 1. We observe that any patch of errors for which $d_{L_1} \leq 2$, i.e., any pattern contained in a cross-shape patch of weight 5, in the bottom part of figure 1 is spread after de-interleaving in a manner that no two or more errors occupy the same row in the top part of figure 1. For example the patch formed by the digits $\Box_3$, $\triangle_4$, $\clubsuit_5$, $\diamondsuit_4$ and $\heartsuit_4$ in the bottom part of figure 1 satisfies $d_{L_1} \leq 2$ and after de-interleaving is spread to five distinct rows. This is an example of perfect spreading.

4 Conclusions

We have introduced *simple interleaving* as a practical technique to correct multiple patches of errors in an $m \times n$ array of digits. We have arbitrarily decided to focus attention on compact patches of errors. This decision turned out to be rewarding as we were able to specify b for

	0	1	2	3	4	5	6
0	$\square_1$	$\square_2$	$\square_3$	$\square_4$	$\square_5$	$\square_6$	$\square_7$
1	$\triangle_1$	$\triangle_2$	$\triangle_3$	$\triangle_4$	$\triangle_5$	$\triangle_6$	$\triangle_7$
2	$\clubsuit_1$	$\clubsuit_2$	$\clubsuit_3$	$\clubsuit_4$	$\clubsuit_5$	$\clubsuit_6$	$\clubsuit_7$
3	$\diamondsuit_1$	$\diamondsuit_2$	$\diamondsuit_3$	$\diamondsuit_4$	$\diamondsuit_5$	$\diamondsuit_6$	$\diamondsuit_7$
4	$\heartsuit_1$	$\heartsuit_2$	$\heartsuit_3$	$\heartsuit_4$	$\heartsuit_5$	$\heartsuit_6$	$\heartsuit_7$

	0	1	2	3	4	5	6
0	$\square_1$	$\triangle_2$	$\clubsuit_3$	$\diamondsuit_4$	$\heartsuit_5$	$\square_6$	$\triangle_7$
1	$\diamondsuit_1$	$\heartsuit_2$	$\square_3$	$\triangle_4$	$\clubsuit_5$	$\diamondsuit_6$	$\heartsuit_7$
2	$\triangle_1$	$\clubsuit_2$	$\diamondsuit_3$	$\heartsuit_4$	$\square_5$	$\triangle_6$	$\clubsuit_7$
3	$\heartsuit_1$	$\square_2$	$\triangle_3$	$\clubsuit_4$	$\diamondsuit_5$	$\heartsuit_6$	$\square_7$
4	$\clubsuit_1$	$\diamondsuit_2$	$\heartsuit_3$	$\square_4$	$\triangle_5$	$\clubsuit_6$	$\diamondsuit_7$

Figure 1: 5×7 array (top) and its interleaved version (bottom).

this class of patches of errors. In ordinary interleaving, used to combat bursts of errors in serial data transmission, an $m \times n$ array contains as rows m codewords of a t random error-correcting code and the array digits are read out by column. Ordinary interleaving allows the correction of at most t bursts of length m per array. The efficiency of *simple interleaving*, measured in terms of the code redundancy employed and the number of errors that can be corrected, is at least the same as that for ordinary interleaving since any t matched patches of errors per array are correctable as are also some other patches of errors. Of course our result in *Theorem 1* can be extended to cover the conditions associated with a perfect interleaving as defined in [2]. However, due to the increase in complexity, we do not feel that to be very important except for dealing with a few cases not fully covered by simple interleaving as, for example, when $m = n = 4$.

References

[1] P.G. Farrell, "An introduction to array error control codes", in *Geometries, Codes and Cryptography*, Editors G. Longo, M. Marchi and A. Sgarro. CISM Courses and Lectures, Springer-Verlag , 1990, pp.101-128.

[2] C. de Almeida and R. Palazzo, Jr.,"Efficient two-dimensional interleaving technique by use of the set partitioning concept." *IEE Electronics Letters*, vol.32, no.6, pp.538-540, 14th March 1996.

[3] A.N. Kauffman, M.A.C. de Moraes, R.C.C. Lima and R.M. Campello de Souza, "A technique for correcting clusters of errors", *International Telecommunication Symposium*, Acapulco, Mexico, 1996.

[4] W.W. Peterson and E.J. Weldon, Jr., *Error-Correcting Codes*, MIT Press, Second Edition, 1972.

[5] H.N. Wright, *First Course in Theory of Numbers*, Dover, New York, 1971.

8. Paddy Farrell and Array Codes

Mario Blaum
IBM Research Division
Almaden Research Center
650 Harry Road
San Jose, CA 95120
blaum@almaden.ibm.com

Abstract

The purpose of this paper is to give a review of the contributions and influence of Prof. Paddy Farrell to the theory and practice of array codes.

1 Introduction

Array codes are two-dimensional (or more generally, multi-dimensional) codes, in which the parity check equations are given by exclusive-OR operations over lines in one or more direction. In this paper only binary codes are considered, although extensions to codes of any symbol size are more or less straightforward.

The purpose of array codes in general is to avoid operations over a finite field (as is the case for Reed-Solomon codes), by using simple exclusive-OR operations, thus reducing the overall complexity and allowing fast encoding and decoding circuits.

Array codes are based on rectangular arrays in which all row-sums and all column-sums (or certain diagonal-sums) have even parity. These arrays define codewords in a block code by reading out the entries in a particular order. However, they may also be viewed as two-dimensional codewords that can be used in applications where information is stored in a two-dimensional medium, like memory chips. When more layers of information are stored on top of each other (for instance by holographic means) one may even have to consider higher-dimensional arrays.

Array codes can also be viewed as a subclass of product codes, composed by all the $n_1 \times n_2$ matrices whose rows belong in some code C_1 and whose columns belong in some code C_2.

We will present different array codes that were deviced or influenced by Paddy Farrell's work. In Section 2, we present array codes for correction of bursts of errors. In Section 3, we give

*This paper is in honor of Paddy Farrell's 60th birthday

a construction of array codes for correction of two-dimensional bursts or *clusters* of errors. In Section 4 we present array codes that are maximum distance separable (MDS) codes in the same sense as Reed-Solomon codes. Finally, in Section 5, we present techniques for 2-dimensional interleaving of error-correcting codes in order to correct 2-dimensional bursts of errors.

In general, proofs will be omitted in this paper and will be referred to the proper references. Let us point out that many of the results presented here can be found in a review article on array codes by the author, together with Paddy Farrell and Henk van Tilborg, that will soon appear in the Handbook of Coding Theory edited by Pless, Huffman and Brualdi [23].

2 Array Codes for Burst Correction

Block *array codes* $\mathcal{A}(n_1, n_2)$ consist of all $n_1 \times n_2$ matrices in which all row-sums and all column-sums have even parity. They can be made into one-dimensional vectors by reading them out in a particular order.

Block array codes $\mathcal{A}(n_1, n_2)$ are linear codes of dimension $(n_1-1)(n_2-1)$ and have minimum Hamming distance 4, as we can easily verify.

To encode information in a block array code $\mathcal{A}(n_1, n_2)$, we first fill an $(n_1 - 1) \times (n_2 - 1)$ array with the information bits in any agreed-upon way. Next we extend all rows with an over-all parity check bit. Finally, extending all n_2 columns with an over-all parity check bit will give an array in $\mathcal{A}(n_1, n_2)$.

Example 2.1 Assume that we want to encode the information vector $\underline{u} = 100110\ 101010$. Putting this vector rowwise into a 2×6 array and extending it in both directions, we obtain the 3×7 array:

1	0	0	1	1	0	1
1	0	1	0	1	0	1
0	0	1	1	0	0	0

Given an array $(a_{i,j})_{\substack{0 \le i \le n_1-1 \\ 0 \le j \le n_2-1}}$, its *horizontal syndrome* $\underline{h} = (h_0, h_1, \ldots, h_{n_1-1})$ and *vertical syndrome* $\underline{v} = (v_0, v_1, \ldots, v_{n_2-1})$ are defined by

$$h_i = \sum_{l=0}^{n_2-1} a_{i,l}, \quad 0 \le i \le n_1 - 1, \tag{1}$$

$$v_j = \sum_{l=0}^{n_1-1} a_{l,j}, \quad 0 \le j \le n_2 - 1. \tag{2}$$

Clearly, an array is in a code $\mathcal{A}(n_1, n_2)$ if and only if both the horizontal and the vertical syndromes are equal to zero.

0	12	3	15	6	18	9
10	1	13	4	16	7	19
20	11	2	14	5	17	8

Figure 1: Readout of code $\mathcal{A}(3,7;2)$

As an example, consider the array

1	0	0	1	1	0	0
1	0	1	0	1	0	1
0	1	1	1	0	0	0

Its horizontal syndrome is $\underline{h} = 1\ 0\ 1$, while its vertical syndrome is $\underline{v} \doteq 0\ 1\ 0\ 0\ 0\ 0\ 1$. It follows that this array is not in $\mathcal{A}(3,6)$.
The error-correcting capability of $\mathcal{A}(n_1,n_2)$ is only one error. This is very poor compared to, say, Hamming codes. However, one of the crucial observations of Paddy Farrell was that, if we change the readout of the bits in $\mathcal{A}(n_1,n_2)$ (say, by reading them diagonally instead of horizontally), burst correction is possible.
More specifically, Farrell and Hopkins [16] discovered that, by reading out the bits diagonally instead of horizontally, burst correction is possible. The burst-correcting capability of the code is a function of n_1, n_2, and the particular diagonal readout used. We shall talk about s-readout if reading starts at the upper left corner of the array and after finishing reading one diagonal, the readout skips $s-1$ diagonals, where s and n_2 are relatively prime, and continues from the top row with the next diagonal. By $\mathcal{A}(n_1,n_2;s)$ we denote the array code with n_1 rows, n_2 columns, and diagonal readout s.
As an example, in Figure 1 the readout for the code $\mathcal{A}(3,7;2)$ is depicted: the first diagonal is followed by the third one, which is followed by the fifth one, and so on. Notice that diagonals are cyclic: after reaching column n_2-1, the next read symbol is in column 0. Also, since the readout is diagonal, bursts will occur along the diagonals.
Since the transpose of an array code is also an array code, we shall assume that $n_2 \geq n_1$.
Specifically, Farrell and Hopkins conjectured that the codes $\mathcal{A}(n_1,n_2;1)$ have burst correcting capability b equal to n_1-1 if and only if $n_2 \geq 2n_1-3$. This result was proven by Blaum, Farrell and van Tilborg [11].
The notation $\langle k \rangle_n$ denotes the unique integer between 0 and $n-1$ that is congruent to k modulo n. For instance, $\langle 8 \rangle_5 = 3$ and $\langle -8 \rangle_5 = 2$.
Given a binary array $(a_{i,j})_{\substack{0\leq i\leq n_1-1\\ 0\leq j\leq n_2-1}}$, the s readout is given by the following rule (where the notation $a \rightarrow b$ means that symbol a is followed by symbol b):

$$a_{i,j} \rightarrow a_{i+1,\langle j+1\rangle_{n_2}}, \quad \text{for } 0 \leq i \leq n_1-2, \tag{3}$$

$$a_{n_1-1,j} \rightarrow a_{0,\langle j-n_1+s+1\rangle_{n_2}}. \tag{4}$$

0	5	10	15	20	25	30
31	1	6	11	16	21	26
27	32	2	7	12	17	22
23	28	33	3	8	13	18
19	24	29	34	4	9	14

Figure 2: Readout of code $\mathcal{A}(5,7;1)$

Note that (3) is obvious and that (4) follows from the observation that $a_{n_1-1,j}$ is the last element on the diagonal starting with $a_{0,\langle j-n_1+1\rangle_{n_2}}$ and so its successor will be $a_{0,\langle j-n_1+1+s\rangle_{n_2}}$. According to this readout, when displayed as a vector with entries $0,1,\ldots,n_1n_2-1$, the 0-th coordinate in the vector corresponds to the entry $a_{0,0}$ in the array, the 1-st location to the entry $a_{1,1}$, and so on. In general,

$$\text{entry } a_{i,j} \text{ corresponds to location } \langle -(n_1-1)i + n_1 j\rangle_{n_1n_2}. \tag{5}$$

Lemma 2.1 The burst error correcting capability b of any $\mathcal{A}(n_1,n_2;s)$ code, $\gcd(s,n_2)=1$, satisfies

$$b \le \begin{cases} n_1-1, & \text{if } s=1, \\ n_1, & \text{if } 1<s<n_2. \end{cases} \tag{6}$$

In [26] W. Zhang and Wolf established an explicit expression for the burst-correcting capability of $\mathcal{A}(n_1,n_2;s)$. Zhang and Wolf's formula is very complex and we omit it here. They also give extensive tables that list, for each $n_1 \le n_2$, $n_2 \le 45$, the largest possible value of the burst correction capability b and the smallest readout s that realizes this.
Part of the tables in [26] were already explained by Farrell and Hopkins result for the case $s=1$. Wanting to explain further the tables in [26], Blaum, Farrell and van Tilborg [12] proved that the codes $\mathcal{A}(n_1,n_2;n_2-1)$ are n_1-burst correcting if and only if $n_2 \ge 2n_1+1$. In that same paper, the authors extended the results to multiple burst correction.
It may seem that the (n_2-1)-readout, allowing for n_1-burst correction, is superior to the 1-readout with (n_1-1)-burst correction. However, this is not necessarily true. Indeed, to minimize the redundancy for b-burst correction when using an $\mathcal{A}(n_1,n_2;1)$ code, the best choice of n_1 and n_2 is $n_1=b+1$, and $n_2=2n_1-3=2b-1$. This code has dimension $(n_1-1)(n_2-1)=2b(b-1)$ and redundancy $n_1+n_2-1=3b-1$. Similarly, if we use an $\mathcal{A}(n_1,n_2;n_2-1)$ code, minimizing the redundancy gives $n_1=b$, $n_2=2n_1+1=2b+1$. This code has the same dimension $(n_1-1)(n_2-1)=2b(b-1)$ while its redundancy is $n_1+n_2-1=3b$. Note that both codes have a redundancy that is about 50% more than the Reiger [2] lower bound of $2b$.
We define the *burst-correcting efficiency* of a code as the quotient between twice its burst-correcting capability and its redundancy, i.e., the efficiency is given by $2b/r$. By the Reiger

bound, this efficiency is at most 1, and we see that it tends to 2/3 for the burst-correcting codes $\mathcal{A}(n_1, 2n_1 - 3; 1)$ and $\mathcal{A}(n_1, 2n_1 + 1; n_1 - 1)$. In [3] Blaum studied families of array codes with $2b/r$ tending to 4/5 using the $s = n_2 - 2$ readout. These results were generalized by Z. Zhang [25], who proved the existence of families of array codes with $2b/r$ tending to 1. Before presenting Zhang's result, we need some further notation.
Denote by $e(n_1, n_2; n_2 - s)$ the efficiency of the code $\mathcal{A}(n_1, n_2; n_2 - s)$. Then, define

$$\begin{aligned} e(n_2; n_2 - s) &= \max_{n_1} e(n_1, n_2; n_2 - s) \\ e(-s) &= \lim_{n_2 \to \infty} \sup e(n_2; n_2 - s) \end{aligned}$$

The main result presented by Zhang is the following:

Theorem 2.1 For any s, consider the codes $\mathcal{A}(n_1, n_2; n_2 - s)$. Let $p(s)$ be the smallest prime such that $\gcd(s, p(s)) = 1$. Then,

$$e(-s) = \frac{2p(s) - 2}{2p(s) - 1}.$$

Let $p_1, p_2, \ldots, p_n, \ldots$ be the sequence of prime numbers in increasing order. A special case of Theorem 2.1 is given by the readouts $n_2 - s_i$, where $s_i = \prod_{j=1}^{i} p_j$. Then we have the following corollary:

Corollary 2.1

$$e(-s_i) = \frac{2p_{i+1} - 2}{2p_{i+1} - 1}.$$

Since $p_i \to \infty$ as $i \to \infty$, Corollary 2.1 shows the existence of array codes $\mathcal{A}(n_1, n_2; n_2 - s)$ with efficiency tending to 1. Sivarajan, McEliece and van Tilborg [24] proved a similar result by using [26].
In particular, for $i = 1, 2, 3, 4$, Corollary 2.1 gives

$$e(-2) = \frac{4}{5}, \quad e(-6) = \frac{8}{9}, \quad e(-30) = \frac{12}{13}, \quad e(-210) = \frac{20}{21}.$$

3 Array Codes for Correction of Two-Dimensional Bursts

In the previous section we have described how (block) array codes can be used to correct bursts in one-dimensional codewords. We shall now consider codes and codewords that have two dimensions. These codes can be used to protect information in two-dimensional data structures that occur in digital communication, storage (e.g. memory chips) and processing systems. A potential new application is in holographic storage.
So, we are now considering the vector space of $n_1 \times n_2$ binary arrays.

Definition 3.1 A *two dimensional burst* of size $b_1 \times b_2$ (also called a *cluster*) is an $n_1 \times n_2$ array, with all its non-zero entries confined to a $b_1 \times b_2$ (and not smaller) subarray. Cyclic bursts are bursts when viewed cyclically mod n_1 and mod n_2.
The $b_1 \times b_2$ submatrix containing the errors is called the *burst pattern.*

Instead of rectangular two-dimensional bursts one may also want to study other shapes like crosses or more circular shaped two-dimensional bursts.
Two-dimensional burst-correcting array codes have been previously studied in the literature. In [20], Imai gives a generalization of Fire codes. Paddy Farrell himself considered two-dimensional burst-correcting array codes in [15]. In [1], Abdel-Ghaffar, McEliece and van Tilborg completely separate the tasks of determining the pattern of the $b_1 \times b_2$ burst and its location. They focus on the minimum redundancy necessary to determine the burst pattern. Interleaving of codes in one dimension to correct bursts can be generalized to multiple dimensions [6]. Also the idea of using codes over extension fields (like Reed-Solomon codes) generalizes quite easily. In effect, let t_1 divide n_1 and t_2 divide n_2. Make a tiling of the $n_1 \times n_2$ array with $t_1 \times t_2$ tiles. Interpret each tile as a symbol from $GF(2^{t_1 t_2})$ and use an error-correcting code over $GF(2^{t_1 t_2})$ to correct symbol errors. Notice that, as in the one dimensional case, a burst in the binary two-dimensional code will cause only a few symbol errors.
For linear $(b_1 \times b_2)$-burst correcting codes, it is best to view its parity check matrix as an $n_1 \times n_2$ matrix whose entries are columns of length r, where r is the redundancy of the code. Clearly two different bursts of size $b_1 \times b_2$ should have different syndromes. This proves:

Lemma 3.1 (Two-dimensional Reiger bound) The redundancy r of a two-dimensional $(b_1 \times b_2)$-burst correcting code satisfies $r \geq 2b_1 b_2$.

Next we will describe a class of two-dimensional $(b_1 \times b_2)$-burst correcting codes $Cl(n_1, n_2; b_1, b_2)$ (Cl stands for cluster) called *staggered* codes. The length of the codes is $n_1 \times n_2$. The encoding and decoding algorithms involve simple exclusive-OR operations and cyclic shifts of binary vectors only. The ideas extend and refine those presented first in [15].
To construct the code we again start with matrices having even parity on rows and on columns. Further, we require the following conditions:

$$n_1 \geq 2b_1 b_2 - b_1 \ , \ n_2 \geq 2b_1 b_2 \ , \ b_1 \text{ divides } n_1 \ , \ b_2 \text{ divides } n_2. \tag{7}$$

The encoding of the staggered code $Cl(n_1, n_2; b_1, b_2)$ is described next.

Algorithm 3.1 (Encoding of $\mathbf{Cl(n_1, n_2; b_1, b_2)}$) Consider an $(n_1 - 1) \times (n_2 - 1)$ array of information.

i) Add a parity bit to each row and to each column of the array such that each row and each column have even parity in the resulting $n_1 \times n_2$ array. This process is depicted in Figure 3 i) for $Cl(10, 12; 2, 3)$. Denote the result by A.

$A =$

1	0	1	1	1	0	1	1	0	0	0	0
1	1	0	0	0	1	1	1	1	0	0	0
0	1	1	0	0	1	1	0	0	1	0	1
0	1	1	1	0	0	1	0	1	1	0	0
0	0	0	0	1	1	1	0	1	0	0	0
1	1	0	1	1	0	0	0	0	0	0	0
0	1	0	1	1	0	0	1	1	1	0	0
1	0	1	1	0	0	0	1	1	1	0	0
1	1	0	1	1	0	0	1	1	0	0	0
1	0	0	0	1	1	1	1	0	0	0	1

i) Adding overall parity check bits to the rows and columns

$A' =$

1	0	1	1	1	0	1	1	0	0	0	0
0	0	0	1	1	0	0	0	1	1	1	1
0	1	1	0	0	1	1	0	0	1	0	1
1	0	0	0	1	1	1	0	0	1	0	1
0	0	0	0	1	1	1	0	1	0	0	0
0	0	0	1	1	0	1	1	0	0	0	0
0	1	0	1	1	0	0	1	1	1	0	0
1	0	0	1	0	1	1	0	0	0	1	1
1	1	0	1	1	0	0	1	1	0	0	0
0	0	1	1	0	0	0	1	1	1	1	0

ii) Row-staggering

$A'' =$

1	1	0	1	1	0	1	1	1	0	0	0
0	0	0	1	0	1	0	1	0	1	1	1
0	0	0	0	1	0	1	1	1	1	0	0
1	0	1	0	1	0	1	0	1	1	1	0
0	1	1	0	0	0	1	0	0	0	0	0
0	0	0	1	1	0	1	0	1	0	0	1
0	0	1	1	1	1	0	0	0	1	0	1
1	0	0	1	1	1	1	1	0	0	0	1
1	1	0	1	1	1	0	1	1	0	0	0
0	0	0	1	0	0	0	0	0	1	1	0

iii) Column-staggering

Figure 3: Encoding of $Cl(10, 12; 2, 3)$ staggered code

ii) Next, for each $i, 0 \leq i \leq n_1 - 1$, cyclically shift row i over $b_2\langle i\rangle_{b_1}$ positions to the right. This means, rows $0, b_1, 2b_1, \ldots$ are left invariant, rows $1, b_1+1, 2b_1+1, \ldots$ are shifted over b_2 positions, etcetera. Denote by A' the array obtained in this way (row-staggering). See Figure 3 ii) where this row-staggering is depicted for $Cl(10, 12; 2, 3)$.

iii) Finally, for each $j, 0 \leq j \leq n_2 - 1$, cyclically shift the j-th columns of A', over $b_1\langle j\rangle_{b_2}$ positions downwards. Denote by A'' the resulting array. See Figure 3 iii) where this column-staggering is depicted for $Cl(10, 12; 2, 3)$.

This completes the encoding.

The array A'' is a staggered version of A, first on rows and then on columns (although the staggering ordering is irrelevant). The set of codewords $Cl(n_1, n_2; b_1, b_2)$ obtained with this encoding procedure is clearly a linear code.

Theorem 3.1 Consider code $Cl(n_1, n_2; b_1, b_2)$, where n_1, n_2, b_1 and b_2 satisfy (7). Then, $Cl(n_1, n_2; b_1, b_2)$ can correct any $b_1 \times b_2$ cluster.

For a proof of Theorem 3.1 together with a decoding algorithm, see [10].
An interesting research problem is finding two-dimensional array codes correcting clusters of size up to $b_1 \times b_2$ with higher efficiency than the staggered codes $Cl(n_1, n_2; b_1, b_2)$.

4 Maximum Distance Separable (MDS) Block Array Codes

In this section we present results in which, although Paddy Farrell is not directly an author, he conjectured some of them in early versions of [15] and in private conversations with the author.
We describe constructions of codes that can correct erased columns or columns in error in a 2-dimensional rectangular array. An application of these results is in multitrack magnetic recording and in RAID type of architectures (i.e., disk arrays) [22].
In general, we will have an $m \times n$ array code, $n - t$ rows carry information and t rows carry redundancy. The code is maximum distance separable (MDS) if and only if up to t erased columns can be corrected, since the Singleton bound [2][21] is met.
The problem of finding MDS block array codes with two parities was solved in [3] and independently in [18][19]. Explicitly, it is proven that an $(m-1) \times m$ array code, with horizontal and diagonal parity lines (with a toroidal topology), can correct any $(m-1)$-bit column in error if and only if m is a prime. The codes can also correct two erased columns, making them very attractive for disk arrays technology. Also, in [18][19] a more general problem is solved, in which the arrays can have any rectangular form, although they are not MDS any more.
We will state the solutions to these problems together with some generalizations.

	0	1	2	3	4
0	0	0	0	0	0
1	1	1	1	1	1
2	2	2	2	2	2
3	3	3	3	3	3
4	4	4	4	4	4

Horizontal parity lines

	0	1	2	3	4
0	0	1	2	3	4
1	1	2	3	4	0
2	2	3	4	0	1
3	3	4	0	1	2
4	4	0	1	2	3

Diagonal parity lines

Figure 4: The two types of parity check lines of the code $\mathcal{B}_2(5)$

Assume that we have m tracks or disks. Let the information in each disk be divided into symbols, where a symbol is a binary vector (bit, byte, sector, etc.). For simplicity, we assume that each symbol is a bit.
The codes we want to construct are defined by means of $(m-1) \times m$ arrays $(a_{i,j})_{\substack{0 \le i \le m-2 \\ 0 \le j \le m-1}}$, m prime, such that $a_{i,j}$ represents the i-th symbol in the j-th column (disk). We will assume that the last two columns (disks) carry the redundancy. However, in order to avoid bottleneck effects when repeated "write" operations are involved, this distribution of the redundancy may be rotated for different blocks in the disk. We shall further add a fictitious zero-row to the array, i.e., $a_{m-1,j} = 0$ for $0 \le j \le m-1$. This is merely done to make the notation easier in the encoding procedure. With this convention, the arrays are now square $m \times m$ arrays. We define the code $\mathcal{B}_2(m)$ as the set of arrays $(a_{i,j})_{0 \le i,j \le m-1}$, with $a_{m-1,j} = 0$ for all j, satisfying the following two equations (recall that $\langle m \rangle_n$ denotes the unique integer l such that $0 \le l < n$ and $l \equiv m \pmod{n}$):

$$\sum_{j=0}^{m-1} a_{i,j} = 0 \ , \quad 0 \le i \le m-2 \tag{8}$$

$$\sum_{l=0}^{m-1} a_{\langle j-l \rangle_m, l} = 0 \ , \quad 0 \le j \le m-1. \tag{9}$$

Equations (8) and (9) define the encoding. There are two types of parity check equations: horizontal and diagonal. The array has the topology of a torus, so the diagonals are wrapped around. The symbols 0, 1, 2, 3 and 4 in Figure 4 depict the horizontal and diagonal parity lines of the code when $m = 5$. Since the fictitious 0-row has been added, we have a 5×5 array.
Code $\mathcal{B}_2(m)$ can trivially correct a column of erased symbols. For that the horizontal parity check symbols would already have been enough. Let us illustrate the decoding of two erased columns with an example.

Example 4.1 Consider the 4×5 array on the left in Figure 5, which is extended with an all-zero row. It contains two erased columns: columns 1 and 4.

	0	1	2	3	4
0	0	?	1	0	?
1	1	?	1	1	?
2	0	?	0	1	?
3	1	?	0	1	?
4	0	0	0	0	0

Before decoding

	0	1	2	3	4
0	0	1	1	0	0
1	1	0	1	1	1
2	0	0	0	1	1
3	1	1	0	1	1
4	0	0	0	0	0

After decoding

Figure 5: An example of decoding two erasures

To decode, start with any of the two columns, say column 1, and consider the diagonal through the known entry $(4,1)$ in this column at the bottom. All elements on this diagonal are known, except where it meets the other erased column. So this entry, which is $a_{1,4}$ can now be computed. It must be 1. With the horizontal parity also $a_{1,1}$ can now be computed. Its value is 0. This concludes the first cycle of the algorithm. So after row 4 was completely known, now also row 1 is completely known. Starting with the diagonal going through position $(1,1)$ we find $a_{3,4} = 1$ (note that $4 - 2.3 \equiv 3 \pmod 5$) and subsequently $a_{3,1} = 1$ (each time the row three higher, modulo 5, will be completely known). Continuing in this way we get $a_{0,4} = 0$ and $a_{0,1} = 1$ and finally $a_{2,4} = 0$ and $a_{2,1} = 1$,

We can now state the main theorem regarding the codes $\mathcal{B}_2(m)$.

Theorem 4.1 Code $\mathcal{B}_2(m)$ can correct up to any two erased columns if and only if m is prime.

The requirement that m is a prime number is not a serious limitation in multitrack magnetic recording and disk arrays applications, since the code can always be shortened to any number of columns by assuming that a certain number of information columns are zero.
Also, a decoding algorithm correcting up to one column in error is not hard to find. For proofs and details, we refer the reader to [3][13][18][19].
The codes described above are MDS but very short: their length is roughly equal to the size of the symbol, while Reed-Solomon codes, also being MDS, are much longer. If the symbols have length m, Reed-Solomon codes have length up to 2^m. So, Reed-Solomon codes can always be used instead of $\mathcal{B}_2(m)$ codes, but the converse is not necessarily true. However, $\mathcal{B}_2(m)$ codes have less complexity, since they involve exclusive-OR operations only, avoiding the arithmetic over a Galois field. There are applications in which the length of the code has size comparable to or smaller than the size of the symbols. Two of these applications are multi-track magnetic recording and redundant arrays of inexpensive disks (RAID) architectures [17][22].
A natural question is how to extend the $\mathcal{B}_2(m)$ codes above to more parities. In effect, the codes $\mathcal{B}_2(m)$ can be naturally extended to codes $\mathcal{B}_r(m)$ with r parities. We define the code

$\mathcal{B}_r(m)$ as the set of arrays $(a_{i,j})$, $0 \leq i,j \leq m-1$, $a_{m-1,j} = 0$ for all j, satisfying the following r equations:

$$\sum_{l=0}^{m-1} a_{\langle j-tl \rangle_m, l} = 0, \quad 0 \leq j \leq m-1, \; 0 \leq t \leq r-1. \tag{10}$$

We can easily see that equations (10) are equivalent to equations (8) and (9) when $r = 2$. For instance when $r = 3$ we get the extra parity check equations

$$\sum_{l=0}^{m-1} a_{\langle j-2l \rangle_m, l} = 0, \quad 0 \leq j \leq m-1,$$

depicted below on the left. On the right an element in $\mathcal{B}_3(5)$ is given.

	0	1	2	3	4
0	0	1	2	3	4
1	3	4	0	1	2
2	1	2	3	4	0
3	4	0	1	2	3
4	2	3	4	0	1

Parity lines of slope 2

1	1	0	0	0
1	0	0	1	0
0	1	1	1	1
1	0	0	1	0
0	0	0	0	0

An array in $\mathcal{B}_3(5)$

The decoding algorithm for r erasures in $\mathcal{B}_r(m)$ when $r > 2$ does not extend trivially from the case $r = 2$. If, for instance, we want to encode $m-3$ columns in $\mathcal{B}_3(m)$ such that the last three columns are redundant, none of the three parity lines intersects only one of those last three columns. Therefore, we cannot apply the method of Example 4.1. In order to solve this problem, we need to use an algebraic description of the codes. Specifically, we need an algebraic description of the $\mathcal{B}_r(m)$ codes as Reed Solomon codes over the ring of polynomials modulo $1 + x + \cdots + x^{m-1}$, m a prime. However, we do not have the space here to do so and we refer the reader to [13].

We have seen that the codes $\mathcal{B}_2(m)$ can correct any two erasures using horizontal and diagonal parity lines. The decoding of two erasures involved a simple recursion. The drawback of the construction is that the processes of encoding and of small write operations also involve a recursion. There are applications, like in RAID architectures, in which the size of each individual symbol can be as big as a whole sector. As a consequence, we will want to minimize the number of redundant symbols affected by the update of a single information symbol. The structure of $\mathcal{B}_2(m)$ forces the updating of most of the redundant symbols each time an information symbol is updated.

In order to overcome this problem, we present an efficient encoding procedure, called EVEN-ODD [5], that is based on exclusive-OR operations and independent parities. Since the parities are independent, there is no recursion during this encoding procedure. The decoding algorithm for two erasures is also simple and similar to the one of $\mathcal{B}_2(m)$. As a result of the

simple encoding procedure a limited write action is greatly simplified, since any modified information symbol affects only two symbols in the redundancy most of the time.
We will assume that the elements of the code are $(m-1) \times (m+2)$ arrays $(a_{i,j})_{\substack{0 \leq i \leq m-2 \\ 0 \leq j \leq m+1}}$, m a prime number, with the information stored in columns 0 to $m-1$ and the redundant data stored in columns m and $m+1$. We also assume, as in the previous sections, that there is an imaginary 0-row after the last row, i.e., $a_{m-1,j} = 0$, $0 \leq j \leq m+1$ (with this convention, the array is now an $m \times (m+2)$ array).
Next we define the code as follows:

Definition 4.1 Given m prime, the code $\mathcal{EO}_2(m)$ is the set of arrays $(a_{i,j})_{\substack{0 \leq i \leq m-2 \\ 0 \leq j \leq m+1}}$, such that for each l, $0 \leq i \leq m-2$,

$$a_{i,m} = \sum_{t=0}^{m-1} a_{i,t} \tag{11}$$

$$a_{i,m+1} = S + \left(\sum_{t=0}^{m-1} a_{\langle i-t \rangle_m, t} \right) \tag{12}$$

where S is defined by

$$S = \sum_{t=1}^{m-1} a_{m-1-t,t}, \tag{13}$$

Equations (11) and (12) define the encoding. We have two types of redundancy: horizontal redundancy and diagonal redundancy. Column m is simply the exclusive-OR of columns $0, 1, \ldots, m-1$ (depicted on the left of Figure 6).
Column $(m+1)$ carries the diagonal redundancy according to Equation (12). Notice that one diagonal is missing in (12). This diagonal consists of the entries $(m-2, 1), (m-3, 2), \ldots, (0, m-1)$ (depicted by ∞ in Figure 6 on the right) and in fact it defines S. So it is this special diagonal that decides the parity of the other diagonals. So if the special diagonal has EVEN parity then so do the others. Otherwise, they all have ODD parity. This is why this scheme is called the EVENODD scheme.
It also follows from the above that

$$\sum_{i=0}^{m-2} a_{i,m} = \sum_{i=0}^{m-2} \sum_{j=0}^{m-1} a_{i,j} = S + \sum_{i=0}^{m-2} a_{i,m+1}.$$

In other words, the parity S of the special diagonal can also be computed as follows:

$$S = \text{the binary sum of the elements in columns } m \text{ and } m+1. \tag{14}$$

Code $\mathcal{EO}_2(m)$ defined above can recover the information lost in any two columns, as the $\mathcal{B}_2(m)$ code previously defined. Therefore, the minimum distance of the code is 3, in the

0	0	0	0	0	0	
1	1	1	1	1	1	
2	2	2	2	2	2	
3	3	3	3	3	3	

Horizontal syndromes

0	1	2	3	∞		0
1	2	3	∞	0		1
2	3	∞	0	1		2
3	∞	0	1	2		3
∞	0	1	2	3		∞

Diagonal syndromes

Figure 6: Parity lines of $\mathcal{EO}_2(5)$

sense that any non-zero array in the code has at least 3 columns that are non-zero. The proof, again, relies on the fact that m is a prime number [5].
The next example illustrates the encoding for $m = 5$.

Example 4.2 Let $m = 5$, and let the symbols be denoted by a_{ij}, $0 \leq i \leq 3$, $0 \leq j \leq 6$. The redundant symbols are in columns 5 and 6. A practical implementation of this example is to consider 7 disks numbered 0 through 6, each disk has 4 disk sectors, the data sectors are on disks numbered 0, 1, 2, 3 and 4, and the redundant disk sectors are on disks numbered 5 and 6. Equation (13) gives

$$S = a_{3,1} + a_{2,2} + a_{1,3} + a_{0,4}.$$

According to Equations (11) and (12) the redundant symbols are obtained as follows:

$$a_{l,5} = a_{l,0} + a_{l,1} + a_{l,2} + a_{l,3} + a_{l,4}\ ,\ 0 \leq l \leq 3$$

$$\begin{aligned} a_{0,6} &= S + a_{0,0} + a_{3,2} + a_{2,3} + a_{1,4} \\ a_{1,6} &= S + a_{1,0} + a_{0,1} + a_{3,3} + a_{2,4} \\ a_{2,6} &= S + a_{2,0} + a_{1,1} + a_{0,2} + a_{3,4} \\ a_{3,6} &= S + a_{3,0} + a_{2,1} + a_{1,2} + a_{0,3} \end{aligned}$$

For instance, assume that we want to encode the 5 columns (we add the imaginary 0-row) below on the left.
We have to fill up the last two columns with the encoded symbols. Notice that $S = a_{3,1} + a_{2,2} + a_{1,3} + a_{0,4} = 1$. Therefore, the diagonals will have odd parity. The encoding gives the array on the right:

1	0	1	1	0		
0	1	1	0	0		
1	1	0	0	0		
0	1	0	1	1		
0	0	0	0	0	0	0

1	0	1	1	0	1	0
0	1	1	0	0	0	0
1	1	0	0	0	0	1
0	1	0	1	1	1	0
0	0	0	0	0	0	0

Let us now give an example illustrating the decoding of the EVENODD code $\mathcal{EO}_2(m)$ for correction of two erasures.

Example 4.3 Consider $\mathcal{EO}_2(5)$, as in Example 4.2, and assume that we have the array below at the left, in which columns (disks) 0 and 2 have been erased (lost):

?	0	?	1	0	1	1
?	1	?	0	0	0	1
?	1	?	0	0	1	1
?	1	?	1	1	0	0
0	0	0	0	0	0	0

0	0	0	1	0	1	1
1	1	0	0	0	0	1
0	1	0	0	0	1	1
1	1	0	1	1	0	0

This is the main case for the decoding algorithm: two information columns have been erased. The cases in which at least one of the two parity columns has been erased are special cases that are easy to handle.

The first step is finding the parity S of the diagonals. This value follows directly from (14). In the array above, we can see that the exclusive-OR of the bits in the two redundant columns is 1. It is not difficult to show that this implies that the diagonals have odd parity. Once S has been established, the decoding proceeds very much like the decoding algorithm for $\mathcal{B}_2(m)$ codes.

Next, the algorithm starts a recursion to retrieve the missing bits $a_{l,0}$ and $a_{l,2}$, $0 \leq l \leq 3$. We first need an entry where we can start. Both entries $(4,0)$ and $(4,2)$ would serve this purpose. Taking the diagonal through $(4,0)$ we see that it only has one unknown: $a_{2,2}$. Since the diagonal has parity $S = 1$, we find $a_{2,2} = 0$. Using the horizontal (even) parity, we obtain $a_{2,0} = 0$. Next, we consider the diagonal going through entry (2,0), etcetera. In this way we find successively: $a_{0,2} = 0$, $a_{0,0} = 0$, $a_{3,2} = 0$, $a_{3,0} = 1$, $a_{1,2} = 0$, $a_{1,0} = 1$. The result is depicted above on the right.

For the general decoding algorithm of EVENODD as well as for proofs, again we refer the reader to [5].

The codes $\mathcal{EO}_2(m)$ can be extended to codes $\mathcal{EO}_r(m)$ that consist of $(m-1) \times (m+r)$ arrays with m information columns and r redundant columns, each redundant column independently encoded. More precisely, given m, we say that the code $\mathcal{EO}_r(m)$ is the set of arrays $(a_{i,j})_{\substack{0 \leq i \leq m-2 \\ 0 \leq j \leq m+r-1}}$, such that, for each l, $0 \leq l \leq m-2$, and for each s, $1 \leq s \leq r-1$,

$$a_{l,m} = \bigoplus_{t=0}^{m-1} a_{l,t} \tag{15}$$

$$a_{l,m+s} = S_s + \left(\sum_{t=0}^{m-1} a_{\langle l-st \rangle_m,t}\right), \tag{16}$$

where

$$S_s = \sum_{t=1}^{m-1} a_{\langle -1-st\rangle_m, t}. \tag{17}$$

Therefore, $\mathcal{EO}_r(m)$ has even parity over horizontal lines as given by Equation (15), and even or odd parity over the lines of slope s, $1 \le s \le r-1$, as given by Equations (16). Bit S_s given by Equation (17), determines the parity of the lines of slope s. It can be proven that if m is prime, then code $\mathcal{EO}_3(m)$ is MDS, i.e., it has minimum (column) distance 4. However, for $r \ge 4$, this is in general not true any more. For details on the generalization of EVENODD, we refer the reader to [8].

5 Two-Dimensional Interleaving Techniques

Although this section does not deal exactly with array codes, it is closely related, since the purpose of 2-dimensional interleaving of error-correcting codes is correcting 2-dimensional bursts, the problem we treated in Section 3.

We present 2-dimensional interleaving schemes, for correcting 2-dimensional bursts (or clusters) of errors, where a cluster of errors in this case will be characterized by its area. The interleaving schemes are based on arrays of integers with the property that every connected component of area t consists of distinct integers. These arrays are called t-interleaved arrays [9]. We also extend the concept of t-interleaved arrays by allowing repetitions within the arrays. Namely, a t-interleaved array with repetition r is an array of integers with the property that in every connected component of area t, every integer is repeated at most r times. Next we formally define those concepts.

Definition 5.1 We say that an element (i,j) in a 2-dimensional array is *connected* to elements $(i+1,j)$, $(i-1,j)$, $(i,j+1)$ and $(i,j-1)$, provided those elements exist.

Definition 5.2 A *path* of length n from E_0 to E_n in a 2-dimensional array is a set of $n+1$ elements $\{E_i \mid 0 \le i \le n\}$ such that for every $0 \le i < n$, element E_i is connected to element E_{i+1}.

Definition 5.3 We say that a set of t elements in a 2-dimensional array is a *cluster* of size t, if any two elements in the cluster belong in a path contained in the set.

The concept of a cluster of size t generalizes in two dimensions the concept of a burst of size t in one dimension. The same idea can be generalized to multiple dimensions (see [9]).

Example 5.1 The 1's in the array below constitute a cluster of size 7.

0	0	0	0	0	0	0
0	1	1	1	0	0	0
0	0	1	1	1	0	0
0	0	0	1	0	0	0

Definition 5.4 Let $t \geq 1$ and $r \geq 1$ be integers. Let $A(t,r)$ be a 2-dimensional array of integers, namely, the elements of the array are labeled by integers. We say that $A(t,r)$ is *t-interleaved* with repetition r if every cluster of size t in $A(t,r)$ consists of integers that repeat at most r times. The *degree of interleaving* of the array is the number of distinct integers it contains.

Notice that, if the integers represent different codes (like in the one-dimensional case), then r-error-correcting codes distributed in a t-interleaved array with repetition r can correct any cluster of size up to t.

Example 5.2 The following is an $A(3,1)$ interleaved array, namely, it is 3-interleaved with repetition 1. Notice that the degree of interleaving is 5.

$A(3,1) =$

0	1	2	3	4	0	1
3	4	0	1	2	3	4
1	2	3	4	0	1	2
4	0	1	2	3	4	0
2	3	4	0	1	2	3

Here we need only five 1-error correcting codes to correct any cluster of size 3. If we are using MDS codes like RS codes, since each code has two redundant symbols, the total redundancy is 10.

Example 5.3 The following is $A(3,2)$, namely, it is 3-interleaved with repetition 2. Notice that the degree of interleaving 2:

$A(3,2) =$

1	0	1	0	1	0	1
0	1	0	1	0	1	0
1	0	1	0	1	0	1
0	1	0	1	0	1	0
1	0	1	0	1	0	1

Here we need only two 2-error correcting codes to correct any cluster of size 3. If we are using MDS codes like RS codes, since each code has four redundant symbols, the total redundancy is 8.

An interesting situation surges from Examples 5.2 and 5.3. Both schemes correct clusters of size 3, however, using 2-error correcting codes gives less total redundancy than 1-error correcting codes. There is no parallel to this when we correct 1-dimensional bursts. Paddy made this interesting observation to the author and to Shuki Bruck, which originated this work.

Our goal is to construct $A(t,r)$ arrays with minimal degree. Notice that in the one-dimensional case, the minimal degree of interleaving $\lceil t/r \rceil$ corresponds linearly to the size of the burst we want to correct. This is not the case in the 2-dimensional case. In [9] optimal degree constructions of $A(t,1)$ arrays for arbitrary t were presented. In [7], the focus was on the case $r=2$. Let us state a lower bound in this case.

Theorem 5.1 Let $t \geq 2$. Let $A(t,2)$ be a t-interleaved array with repetition 2. Then

1. If t is even, then the degree of interleaving of $A(t,2)$ is at least $\frac{(t/2)(1+t/2)}{2}$.
2. If t is odd, then the degree of interleaving of $A(t,2)$ is at least $\lceil\frac{(t+1)^2}{8}\rceil$.

The idea in the proof is to construct arrays of integers of size $\frac{(t/2)(1+t/2)}{2}$ in the case t even, and size $\lceil\frac{(t+1)^2}{4}\rceil$, in the case of t odd, and prove that if those arrays are part of $A(t,2)$ then every integer in the array is repeated at most twice.
Let us illustrate the idea of the proof by an example. Consider the following 2 by 3 array of labels called $B(4)$.

$B(4) =$

a	b	c
d	e	f

Note that that every three labels in $B(4)$ are part of a cluster of size 4. This follows from the fact that for every three labels, two are in the same row. Namely, a cluster that contains this row and the third label is of size 4. For example, let a, c and e be the three labels then the cluster that contains those labels is a, b, c and e.
Let us end this section by discussing constructions of $A(2,t)$ interleaved arrays. First we describe an interleaving scheme that we call the *toroidal* interleaving scheme.

Construction 5.1 Consider a 2-dimensional array and an integer m. Label the coordinates of the array toroidally on m, i.e., the coordinates are given by (x,y), where x and y are taken modulo m. Let b be relatively prime with m. Then, for each a modulo m, the coordinates $(i, a+ib)$ (taken modulo m) are assigned the same number a.

Example 5.4 Assume that we have a 4×6 array. Taking $m=2$ and $b=1$, Construction 5.1 gives the following $A(3,2)$ interleaved array:

$A(3,2) =$

0	1	0	1	0	1
1	0	1	0	1	0
0	1	0	1	0	1
1	0	1	0	1	0

Similarly, if we consider a 5×10 array for $m=5$ and $b=3$, we obtain the following $A(5,2)$ interleaved array:

$A(5,2) =$

0	1	2	3	4	0	1	2	3	4
2	3	4	0	1	2	3	4	0	1
4	0	1	2	3	4	0	1	2	3
1	2	3	4	0	1	2	3	4	0
3	4	0	1	2	3	4	0	1	2

The reader can verify that the array above is 5-interleaved with repetition 2.

As we can see in Example 5.4, given an array labeled by Construction 5.1, in order to find if the array is t interleaved, it is enough to consider the $m \times m$ array obtained by the construction. The labeling of the whole array is obtained by tiling it with the $m \times m$ array.

Definition 5.5 The *Lee distance* between two elements in a torus is the length of the shortest path they belong to (for example, two adjacent elements are at Lee distance 1). The minimum Lee distance of a set of elements is the minimum of the Lee distance between all the pairs of elements in the set.

The following theorem (proof omitted) gives a method for finding t in Construction 5.1.

Theorem 5.2 Consider Construction 5.1 with parameters m and b. Let d be the minimum Lee distance in the $m \times m$ torus between two coordinates labeled with the same number. Then,

1. For d even, the construction provides an $A(t, 2)$ interleaved array with $t = 3d/2$.

2. For d odd, the construction provides an $A(t, 2)$ interleaved array with $t = (3d + 1)/2$.

The following theorem from [9] describes an optimal method for constructing arrays with a given minimum Lee distance d. Those arrays are in fact $A(d, 1)$ interleaved arrays.

Theorem 5.3 Using construction 5.1 with parameters m and b provides constructions of arrays with minimum Lee distance d between equal labels. Where

1. For d an odd integer, $m = \frac{d^2+1}{2}$, and $b = d$.

2. For d an even integer, $m = \frac{d^2}{2}$, and $b = d - 1$.

Corollary 5.1 Using construction 5.1 with parameters m and b provide constructions of $A(t, 2)$ interleaved arrays. Where,

1. For $t = (3d + 1)/2$, d an odd integer, $m = \frac{d^2+1}{2}$, and $b = d$. Resulting in a degree of interleaving of $\frac{2t^2-2t+5}{9}$.

2. For $t = 3d/2$, d an even integer, $m = \frac{d^2}{2}$, and $b = d - 1$. Resulting in a degree of interleaving of $\frac{2t^2}{9}$.

Example 5.5 Consider the case $d = 4$. According to Corollary 5.1 $m = 8$ and $b = 3$. Therefore, tiling an array with the following 8×8 array gives an $A(6, 2)$ interleaved array:

0	1	2	3	4	5	6	7
5	6	7	0	1	2	3	4
2	3	4	5	6	7	0	1
7	0	1	2	3	4	5	6
4	5	6	7	0	1	2	3
1	2	3	4	5	6	7	0
6	7	0	1	2	3	4	5
3	4	5	6	7	0	1	2

For $d = 5$, according to Corollary 5.1 $m = 13$ and $b = 5$. Therefore, tiling an array with the following 13×13 array gives a $A(8, 2)$ interleaved array:

0	1	2	3	4	5	6	7	8	9	10	11	12
8	9	10	11	12	0	1	2	3	4	5	6	7
3	4	5	6	7	8	9	10	11	12	0	1	2
11	12	0	1	2	3	4	5	6	7	8	9	10
6	7	8	9	10	11	12	0	1	2	3	4	5
1	2	3	4	5	6	7	8	9	10	11	12	0
9	10	11	12	0	1	2	3	4	5	6	7	8
4	5	6	7	8	9	10	11	12	0	1	2	3
12	0	1	2	3	4	5	6	7	8	9	10	11
7	8	9	10	11	12	0	1	2	3	4	5	6
2	3	4	5	6	7	8	9	10	11	12	0	1
10	11	12	0	1	2	3	4	5	6	7	8	9
5	6	7	8	9	10	11	12	0	1	2	3	4

We note that the constructions above are not optimal with respect to the degree of interleaving. In fact, we can improve the degree of interleaving obtained with construction 5.1 by using a different set of parameters. For example, we can construct $A(6, 2)$ with degree 7, compared to degree 8 above, by using $m = 7$ and $b = 2$, as follows,

0	1	2	3	4	5	6
5	6	0	1	2	3	4
3	4	5	6	0	1	2
1	2	3	4	5	6	0
6	0	1	2	3	4	5
4	5	6	0	1	2	3
2	3	4	5	6	0	1

Using computer search we can find optimal sets of parameters for the torodial construction. Table 1 gives lower and upper bounds on the degree of interleaving.

t	Lower bound	Upper bound Optimization
3	2	2
4	3	3
5	5	5
6	6	7
7	8	10
8	10	12(1,5)
9	13	16(2,3)
10	15	19(1,7)
11	18	22(2,3)
12	21	27(1,8)
13	25	33(1,9)
14	28	37(1,10)
15	32	44(1,10)
16	36	48(1,11)
17	41	56(4,5)
18	45	61(1,13)
19	50	69(1,19)
20	55	75(1,14)

Table 1: Lower and upper bounds on the degree of interleaving of $A(t, 2)$.

Acknowledgement: I want to thank Paddy for having introduced me to Array Codes when I was still a graduate student, and for all the fun we had all these years. I cannot believe that Paddy has turned 60. It looks like yesterday when he was waiting for me at the train station in Manchester. That was nearly 14 years ago.
I also want to thank my co-authors in my work on array codes, mainly Shuki Bruck, Ronny Roth, Henk van Tilborg and Alex Vardy. I get by with a little help from my friends.

References

[1] K. A. S. Abdel-Ghaffar, R. J. McEliece, H. C. A. van Tilborg, *Two-dimensional burst-identification and their use in burst correction, IEEE Trans. Inform. Theory,* IT-34 (1988), 494–504.

[2] R.E. Blahut, *Theory and Practice of Error Control Codes,* Addison-Wesley, Reading, Massachusetts, 1983.

[3] M. Blaum, *A family of efficient burst-correcting* array codes, *IEEE Trans. Inform. Theory,* IT-36 (1990), 671–675.

[4] M. Blaum, *A class of byte-correcting* array *codes,* IBM Research Report RJ 5652 (57151), 1987.

[5] M. Blaum, J. Brady, J. Bruck, J. Menon, *EVENODD: An Efficient Scheme for Tolerating Double Disk Failures in RAID* Architectures, *IEEE Trans. on Computers,* Vol. C-44 (1995), 192–202.

[6] M. Blaum, J. Bruck, *Correcting two-dimensional clusters by interleaving of symbols, Proceedings IEEE International Symposium on Information Theory,* Trondheim, Norway, 1994.

[7] M. Blaum, J. Bruck, P. G. Farrell, *Two-Dimensional Interleaving Schemes with Repetitions, Proceedings IEEE International Symposium on Information Theory,* Ulm, Germany, 1997.

[8] M. Blaum, J. Bruck, A. Vardy, *MDS Array Codes with Independent Parity Symbols, IEEE Trans. Inform. Theory,* IT-42 (1996), 529–42.

[9] M. Blaum, J. Bruck, A. Vardy, *Interleaving Schemes for Multidimensional Cluster Errors,* to appear in *IEEE Trans. Inform. Theory.*

[10] M. Blaum, P.G. Farrell, *Array Codes for Cluster-Error Correction, Electronic Letters,* Vol. 30, No. 21, 1752–3, Oct. 1994.

[11] M. Blaum, P.G. Farrell, H.C.A. van Tilborg, *A class of burst* error-correcting array *codes, IEEE Trans. Inform. Theory,* IT-32 (1986), 836–839.

[12] M. Blaum, P.G. Farrell, H.C.A. van Tilborg, *Multiple burst-correcting* array *codes, IEEE Trans. Inform. Theory,* IT-34 (1988), 1061–1066.

[13] M. Blaum, R.M. Roth, *New Array Codes for Multiple Phased Burst Correction, IEEE Trans. Inform. Theory,* IT-39 (1993), 66–77.

[14] P.G. Farrell, *A survey of* array *error control codes,* Vol. 3, No. 5 (1992), 441–454.

[15] P.G. Farrell, *Array codes for correcting cluster error patterns, IEE Conf. on Electronic Signal Processing,* York, 1982.

[16] P.G. Farrell, S.J. Hopkins, *Burst-error-correcting* array codes, *Radio Elec. Engr.*, 52 (1982), 188–192.

[17] M. B. Friedman, *RAID keeps going and going and..., IEEE Spectrum,* (April 1996), 73–79.

[18] R. Goodman, M. Sayano, *Size limits on phased burst error correcting array codes, Elec. Lett.*, 26 (1990), 55–56.

[19] R. Goodman, R.J. McEliece, M. Sayano, *Phased burst error correcting codes, IEEE Trans. on Information Theory,* IT-39 (1993), 684–693.

[20] H. Imai, *Two-Dimensional Fire Codes, IEEE Trans. Inform. Theory,* IT-19 (1973), 796–806.

[21] F.J. MacWilliams, N.J.A. Sloane, *The Theory of Error-Correcting Codes,* North Holland, Amsterdam, 1977.

[22] D. A. Patterson, G. A. Gibson, R. Katz, *A Case for Redundant Arrays of Inexpensive Disks, SIGMOD International Conference on Data Management,* Chicago (1988), 109–116.

[23] V. Pless, C. Huffman, R. Brualdi, *Handbook of Coding Theory,* to appear.

[24] K. N. Sivarajan, R. J. McEliece, H. C. A. van Tilborg, *Burst error-correcting and detecting codes, IEEE Int. Symp. Inform. Theory,* San Diego, CA, 1990.

[25] Z. Zhang, *Limiting efficiencies of burst-correcting array codes, IEEE Trans. Inform. Theory,* IT-37 (1991), 976–982.

[26] W. Zhang, J.K. Wolf, *A class of binary burst error-correcting quasi-cyclic codes, IEEE Trans. Inform. Theory,* IT-34 (1988), 463–480.

9. Array Codes and Cryptography

R. M. Campello de Souza, J. C. de Souza & A. N. Kauffman
Communications Research Group - CODEC - Departamento
de Eletrônica e Sistemas - UFPE, CP 7800, 50732-970
Recife - PE Brasil E-mail Ricardo@npd.ufpe.br

Abstract This paper focuses on a class of multilevel array codes that are capable of correcting a special type of error which we call bi-separable. The codes are single random error-correcting and have the capacity for correcting bi-separable errors of arbitrary weight. Aspects of the encoding and decoding operations are discussed and, as an interesting application, a new private-key cryptosystem is introduced. The capacity of the cryptosystem to resist the attacks that are typically used against encryption techniques based on error control codes is analysed and some considerations concerning the choice of permutations to be used in the scheme are made.

I. Introduction

Array codes are constructed from linear block or convolutional codes which are assembled in two or more geometrical dimensions [1]. They have been studied intensively in the literature and represent a very rich source of many powerful and practical constructions for a number of applications which include, but are not limited to, data transmission, digital audio, magnetic recording, semiconductor memories, compact discs, sattelite communications and cryptography. In most applications emphasis is put on the use of simple constituent codes and low decoding complexity [2].

The use of array codes for cryptographic purposes was firstly considered in [3], where array codes for burst error correction were used to conceive a new type of private-key cryptosystem. Array codes, binary or multilevel, constructed from block or convolutional codes, seem to have a very important role to play in the field of data security. Specifically, as it is shown in this paper, array codes constructed from rather simple constituent codes, which in turn have very simple encoding and decoding algorithms, can be used to conceive computationally secure cryptosystems.

Cryptographic techniques based on error control codes were first introduced by McEliece [4]. Later, Rao and Nam proposed a private-key version of McEliece's system which allowed the use of non-algebraic codes [5]. The ideas introduced in [6] represent a new approach to the problem in the sense that the use of burst-correcting codes for private-key encryption was introduced. In this context, the cryptosystem security relies not on the computational complexity of decoding a code that is only linear, but on the difficulty of correcting a number of errors that is beyond the error-correcting capacity of a given code. To implement such a criptosystem, as described in [3], the array codes introduced in [7] were used.

In this paper, a class of multilevel array codes is investigated aiming its use as part of a private-key encryption scheme. These are array codes defined over $GF(q)$, with random error-correcting capacity $t = 1$ and a capacity to correct a special type of error, called bi-separable, that is a function of q and of the code dimensions and thus may assume an arbitrary value. In the next section, the concept of bi-separable error is introduced and a class of two-dimensional (2-D) array codes for correcting them is presented. The description of a private-key cryptosystem based on these codes, which includes some attempts to cryptanalyse it, is made in section III.

Next, an analysis concerning the choice of permutations suitable to be used with the scheme is presented. Finally, some concluding remarks are made in section V.

II. Bi-Separable Error-Correcting Codes

In general, error control codes based cryptosystems rely on the modification of a code structure, which usually preserves only its linearity, therefore making the task of correcting the errors that are artificially introduced at the transmitter as part of the encryption process, a difficult one. In this context it is clear that it is not required from the error structure to be used, to exist in any real channel and, in principle, any such structure, as long as it produces a computationally secure encryption scheme, may be used. In what follows, a new type of error pattern is defined and a class of array codes to correct it is presented.

Definition 1 - The direct mapping with parameters r and s, $DM(r, s)$, associated to the vector $v = (v_1, v_2, \ldots\ldots, v_{rs})$, maps the rs components of v into the matrix

$$V = \begin{bmatrix} v_1 & v_2 & \cdots & v_s \\ v_{s+1} & v_{s+2} & \cdots & v_{2s} \\ \vdots & \vdots & \ddots & \vdots \\ v_{(r-1)s+1} & v_{(r-1)s+2} & \cdots & v_{rs} \end{bmatrix}$$

and the inverse direct mapping, $IDM(r,s)$, maps the elements of the matrix V back into the vector v .

Definition 2 - Let $e = (e_1, e_2, \ldots, e_{rs})$ be a vector of length rs whose components belong to $GF(q)$. We call e a bi-separable error vector of dimensions $r \times s$ over $GF(q)$, denoting it by $BE(r, s)$, if the following conditions are satisfied :

i - Its nonzero components are distincts elements of $GF(q)$.

ii - The matrix V obtained by applying $DM(r, s)$ to e contains, at most, one nonzero component per row and column.

From (i) it is clear that the maximum number of nonzero components of a $BE(r,s)$ is $q-1$ and, from (ii), this number can be no greater than $min\ (r, s)$. Therefore the maximum weight of a $BE(r, s)$ over $GF(q)$ is $w_{máx} = min\ (q-1, min\ (r,s))$. To correct such error patterns, we use a 2-D array code, whose row and column codes are single parity-check codes over $GF(q)$, $C_1(n_1, k_1, d_1)$ and $C_2(n_2, k_2, d_2)$ respectively, of parameters $n_1 = s$, $k_1 = s-1$, $d_1 = 2$ and $n_2 = r$, $k_2 = r-1$, $d_2 = 2$. Encoding and decoding for this bi-separable error-correcting code (BECC) are as follows.

Encoding - The information positions of the array are filled with the $k = (r-1)(s-1)$ message symbols from which the parity-check symbols are computed. The array is then mapped into a vector of length rs, via $IDM(r, s)$, for transmission.

Decoding - The possibly corrupted symbols of the received vector are mapped back into array form via $MD(r, s)$, and the horizontal and vertical syndromes, denoted respectively by $h = (h_1, h_2, \ldots, h_r)$ and $v = (v_1, v_2, \ldots, v_s)$, are computed. Suposing a $BE(r, s)$ of weight $w \leq w_{máx}$ ocurred, we denote the values of its nonzero components by $a_1, a_2, \ldots, a_w$. Each such component, after the use of $MD(r, s)$, will be occupying different rows and columns. Therefore, both the horizontal and vertical syndromes will have exactly w nonzero components, whose values are $a_1, a_2, \ldots, a_w$, so that the error symbol of value a_x is in row i and column j of the array, where i and j are such that $h_i = v_j = a_x$, $x = 1, 2, \ldots.,$ w. The error is then corrected and decoding is complete. The computational complexity of the above procedures is $O(k)$ sums in $GF(q)$.

The BECC just described belongs to a class of equivalent codes having $L=[(r-1)(s-1)]!rs$ members, corresponding to the different orders of feeding the $k = (r-1)(s-1)$ information symbols in the array and the different locations for the parity-check row and column. Therefore each generator matrix may be described simply by a permutation of degree k and two integers i and j, where $1 \le i \le r$, $1 \le j \le s$. To find the number N_B of codes that are combinatorially equivalent to a given member of the class, we observe that, to preserve the parity-check equations, we may only permutate rows and columns and therefore $N_B = r!s!$.

III. A BECC Based Private-Key Cryptosystem

To construct a private-key criptosystem based on the class of BECC's introduced in the previous section, we proceed as follows. The plaintext m is a q-ary k-tuple, the ciphertext c is a q-ary n-tuple, and they are related by

$$c = (mSG + e_w)P \tag{1}$$

where $G(k \times n)$ is the code generator matrix, e_w is a bi-separable error of weight w, where $1 << w_{min} \le w \le w_{max}$, $S(k \times k)$ is a scrambling matrix and $P(n \times n)$ is a permutation matrix. Deciphering is implemented by firstly multiplying c by P^{-1} and then using the decoding algorithm described previously to get mS, from which m can be recovered after multiplication by S^{-1}. To break the system, the standard approach is to rewrite (1) as

$$c = mG' + e'_w \tag{2}$$

where $G' = SGP$ and $e'_w = e_w P$. At this point it is possible to find G' by a chosen plaintext attack [5]. However, decoding through G' is unfeasible because the code generated by G' can correct only a single error $(t = 1)$ and e'_w, a permuted version of the bi-separable error randomly introduced at the transmitter, has weight $w >> 1$. Therefore, the cryptanalist has to find S, G and P by factoring G' or else try to recover m directly from c without the knowledge of these matrices. In the first case, the important factor is the number N_B of combinatorially equivalent codes to the BECC generated by G, since the cryptanalyst must find a matrix that can be used for decoding, among all possible generator matrices. The probability of success is N_B/L, which implies an average number A of trials given by

$$A = [(r-1)(s-1)]! \,/\, (r-1)!(s-1)! \tag{3}$$

To validate a solution, a known plaintext/ciphertext pair (m, c) must be examined, so that the working factor for this attack is $T_1 = A$ * (deciphering complexity).

To implement the second alternative, subsets of k symbols are randomly chosen from c and tested exhaustively for error patterns of weights up to a given value, say j [8]. The probability of occurrence of up to j errors in the k coordinates, P_j, is given by

$$P_j = \sum_{i=0}^{j} \sum_{w=i}^{w_{max}} \frac{\binom{k}{i}\binom{n-k}{w-i}}{\binom{n}{w}} * \frac{N_{be}(w)}{N_{tbe}} \tag{4}$$

where $N_{be}(w)$ and N_{tbe} are, respectively, the number of bi-separable errors $BE(r,s)$ of weight w and the total number of such errors for a given w_{min}, that is,

$$N_{be}(w) = \binom{q-1}{w} \prod_{i=1}^{w} (r+1-i)(s+1-i) \tag{5}$$

and

$$N_{tbe} = \sum_{w=w_{min}}^{w_{max}} N_{be}(w) \tag{6}$$

The expected number of choices is P_j^{-1} and, for each one of them, N_j tests must be made, where

$$N_j = \sum_{i=0}^{j} \binom{k}{i} (q-1)^i \tag{7}$$

The working factor for this attack is $T_2 = P_j^{-1}(k^{\alpha} + N_j k^{\beta})$, where $2<\alpha<3$ *and* $1<\beta\leq 2$ (the factors α and β are associated, respectively, to the solution of linear systems with k equations and to the inversion of $k \times k$ matrices). For the values $r = 18$, $s = 21$, $q = 32$ and $w_{min} = 19$, *results* $T_2 = 1{,}4 \times 10^{28}$.

IV. On the Choice of Permutations

In many cryptosystems, permutations play an important role and represent a substantial contribution to increase the confusion and diffusion of the encryption process. In the search for cryptographically strong permutations to be used in the cryptosystem of section III, we introduce the arithmetic function $P_n(i)$.

Definition 3 - $P_n(i)$ denotes the number of permutations of degree n that fix i elements, $0 \leq i \leq n$, $n \geq 1$.

Lemma 1 - $P_n(i) = \binom{n}{i} P_{n-i}(0)$

Proof - There are $\binom{n}{i}$ ways of choosing the i elements. For each one, all the remaining $(n-i)$ elements must permute with each other, which, by definition, can be made in $P_{n-i}(0)$ ways.

□

We now proceed to completely specify $P_n(i)$. From lemma 1, we must first to determine $P_{n-i}(0)$ or, in general, $P_n(0)$. As a matter of fact, from the point of view of cryptography, clearly the most relevant $P_n(i)$'s are those for very low values of i.

Lemma 2 - $P_n(0)$ satisfies the recurrence relation

$$P_n(0) = (n-1)[P_{n-1}(0) + P_{n-2}(0)] \quad (8)$$

with initial conditions $P_1(0) = 0$ and $P_2(0) = 1$.

Proof - Clearly, the initial conditions are satisfied. Without loss of generality, we represent the elements to be permuted by the coordinates $1, 2, \ldots\ldots, n$. Focusing our attention on two such elements, say 1 and j, $j = 2, 3 \ldots., n$, there are two situations to be taken into account :
i) 1 and j swap places and stay fixed thereafter. Since there are $(n-1)$ choices for j and $(n-2)$ elements remain to be permuted, the contribution here is $(n-1)P_{n-2}(0)$.
ii) Now 1 goes into j, but not vice-versa. Then it is as if we were starting from (i) after the transposition of 1 and j. Again there are $(n-1)$ starting points for j, but now $(n-1)$ elements remain to be (all) permuted. That contributes with $(n-1)P_{n-1}(0)$. Adding the contributions in (i) and (ii), we obtain the desired relation.

□

We are now in a position to state the main result of the section, namely, a closed formula for $P_n(i)$.

Theorem 1 - $P_n(i) = \frac{n!}{i!}\sum_{j=0}^{n-i}\frac{(-1)^j}{j!}$, $i = 0, 1, \ldots., n$.

Proof - By straightforward substitution (or by induction) it can be readily shown that the solution of the recurrence relation in lemma 2 is

$$P_n(0) = n!\sum_{j=0}^{n}\frac{(-1)^j}{j!}$$

From lemma 1, this implies that

$$P_n(i) = \binom{n}{i}(n-i)!\sum_{j=0}^{n-i}\frac{(-1)^j}{j!}$$

and the result follows after the expansion of the binomial coefficient $\binom{n}{i}$. □

Of special interest to cryptography is the behaviour of $P_n(i)$ for $i = 0$. For security reasons it is desirable that we use, in the encryption process, one of the $P_n(0)$ permutations that fix no elements. However, that can only be done in a way that is both safe and practical, if we have a clear understanding of the relative behaviour of $P_n(0)$ and $n!$. Corollary 1 belows shows an interesting result concerning the assymptotic relative behaviour of these functions.

Corollary 1 - $\lim_{n\to\infty}\left(\frac{n!}{P_n(i)}\right) = i!e$

Proof - The Maclaurin's series expansion of the exponential function $f(x) = e^{-x}$ is

$$e^{-x} = \sum_{j=0}^{\infty} \frac{(-1)^j x^j}{j!}$$

From theorem 1

$$\frac{P_n(i)}{n!} = \frac{1}{i!} \sum_{j=0}^{n-i} \frac{(-1)^j}{j!}$$

so that making $x = 1$ and taking the limit as $n \rightarrow \infty$, we get

$$\lim_{n \to \infty} \left(\frac{P_n(i)}{n!} \right) = \frac{e^{-1}}{i!}$$

and the result follows.

□

Corollary 1 provides some useful indications as to the viability of using one of the $P_n(0)$ permutations. Specifically, it shows that, for large n, $P_n(0)$ and $P_n(1)$ accounts, each one, for aproximately 36.7% of all $n!$ permutations of degree n. Considering the cryptanalyst resources and also practical aspects of implementation, this information can be used in the design of the cryptosystem presented in this paper as well as in any cryptosystem that makes use of permutations as part of its secret key.

V. Conclusions

Array codes are a very important class of error control codes and have been used in many applications. This paper discusses its use for cryptographic purposes. A new type of error pattern , called bi-separable, has been introduced and a class of multilevel 2-D array codes to correct it was presented. As it usually happens with most applications involving array codes, the encoding and decoding (correction of bi-separable error patterns) operations have very low computational complexity. These elements were then applied to the conception of a new private-key cryptosystem. The main attacks that are typically applied to break cryptosystems based on error control codes were analysed and, due to the large values of some parameters concerning the array codes, namely, the total number of bi-separable error patterns and of generator matrices, the computed working factors indicate that the scheme seems able to resist these attacks. We also analysed aspects related to the choice of permutations to be used in the encryption process and some interesting results were derived, which have implications for the cryptosystem design, concerning the number of permutations of degree n that fix a given number i of coordinates.

Acknowledgements

This work received partial support from *Conselho Nacional de Desenvolvimento Científico e Tecnológico - CNPq*, under grant No. 304248/84-3.

References

[1] P. G. Farrell, *An Introduction to Array Error Control Codes.* In : *Geometrics, Codes and Cryptography*, Eds. Longo, Marchi and Sagro, Springer-Verlag, CISM No. 313, 1990.

[2] P. G. Farrell, *A Survey of Array Error Control Codes*, European Transactions on Telecommunications, Vol. 3, No. 5, Sept-Oct, 1992.

[3] R.M. Campello de Souza, J. Campello de Souza, *Array Codes For Private-Key Encryption*, Electronics Letters, Vol. 30, Nº 17, pp. 1394-1396, August 1994.

[4] R.J. McEliece, *A Public-Key Cryptosystem Based on Algebraic Coding Theory*, *DSN Progress Report* 42-44, pp. 114-116, Jet Propulsion Laboratory, Pasadena, CA, Jan./Feb. 1978.

[5] T.R.N. Rao, K. Nam, *Private-Key Algebraic-Code Encryption*, IEEE Transactions, Vol. IT-35, Nº 4, pp. 829-833, July 1989.

[6] F.M.R. Alencar, A.M.P. Léo, R.M. Campello de Souza, *Private-Key Burst Correcting Code Encryption*, Proceedings of the IEEE International Symposium on Information Theory, pp. 227, Jan. 1993.

[7] P. G. Farrell, S. J. Hopkins, *Burst-error-correcting Array Codes*, Radio and Electronic Eng., Vol. 52, No. 4, pp. 182-192, April 1982.

[8] P. J. Lee, F. Brickell, *An Observation on the Security of McEliece's Public-key Cryptosystem.* In : *Advances in Cryptology,* Eurocrypt 88 Proceedings, pp. 275-280, Springer-Verlag, 1988.

CHAPTER 3

Decoding of Error Control Codes

To be presented at a conference in honor of P. G. Farrell's 60th birthday, Lancaster, England, January 26–28, 1998.

10. How to Compute Weight Enumerators for Convolutional Codes

Robert J. McEliece*

Department of Electrical Engineering
California Institute of Technology
Pasadena, California 91103

rjm@systems.caltech.edu

Abstract.

Techniques for finding transfer functions for convolutional codes have long been known, but the methods that have appeared in print are, for the most part, more art than science. In this tutorial paper, we describe a precise and efficient algebraic method called the transfer matrix method. The transfer matrix method allows us not only to compute transfer functions, but also the weight enumerators for certain block codes which can be derived from the parent convolutional code, viz., the sequence of truncated and tailbiting codes. The transfer matrix method is easiest to apply if one uses a symbolic manipulation program like Mathematica, but even so, it is practical only for codes with relatively small degree.

* This work was supported by NSF grant no. NCR-9505975, AFOSR grant no. 5F49620-97-1-0313, and grant from Qualcomm, Inc.

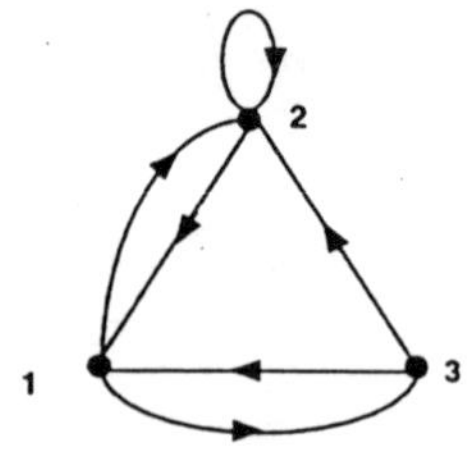

Figure 1. A simple digraph.

1. Digraphs and Trellises. The Transfer Matrix method.

In this section, we summarize the basic combinatorial and algebraic tools we will need to compute the weight enumerators described in Section 2. Much of the material in this section is taken from Stanley [1, Section 4.7].

A finite *directed graph* or *digraph* D consists of a finite vertex set V and a finite edge set E, where $E \subseteq V \times V$. If $e = (\alpha, \beta) \in E$, then e is said to be a *directed edge* from α to β, and to have *initial vertex* α, denoted $\mathrm{init}(e) = \alpha$, and *final vertex* β, denoted $\mathrm{fin}(e) = \beta$. Figure 1 shows a simple digraph with vertex set $V = \{1, 2, 3\}$ and edge set $E = \{(1,2), (1,3), (2,1), (2,2), (3,1), (3,2)\}$.

A *path of length K from u to v* in D is a sequence of K edges $e_1 e_2 \cdots e_K$, such that $\mathrm{init}(e_1) = u$, $\mathrm{fin}(e_K) = v$, and $\mathrm{fin}(e_i) = \mathrm{init}(e_{i+1})$, for $1 \leq i < K$. For example, in the digraph of Figure 1, $((1,2), (2,2), (2,1), (1,3))$ is a path of length 4 from 1 to 3. A path of length K can also be represented by a sequence of $K+1$ vertices $v_0 v_1 \cdots v_K$, where $(v_{i-1}, v_i) = e_i$. For example the path $((1,2), (2,2), (2,1), (1,3))$ in Figure 1 can also be represented by the vertex sequence $(1, 2, 2, 1, 3)$.

Now let $w : E \mapsto R$ be a *weight function* on the edge set E with values in some commutative semiring R. If $P = e_1 e_2 \cdots e_K$ is a path, then the *weight* of P is defined to be $W(P) = w(e_1) w(e_2) \cdots w(e_K)$. For example, in Figure 1, if we assume that all edges have weight 1, the path $((1,2), (2,2), (2,1), (1,3))$ also has weight 1.

A digraph D with a weight function on its edges can be represented a square matrix A, called the *incidence matrix* for D, whose rows and columns are indexed by V. If $e = (\alpha, \beta) \in E$, then the (α, β) entry of A is equal to $w(e)$.[1] If there is no edge from α to β, the (α, β) entry of A is equal to 0. For example, the digraph of Figure 1 can be

[1] If we allow multiple edges between pairs of vertices, the corresponding matrix entry is defined to be the sum of the weights of all such edges.

represented by the incidence matrix

$$(1.1) \qquad A = \begin{array}{c} \\ 1 \\ 2 \\ 3 \end{array} \begin{array}{c} \begin{array}{ccc} 1 & 2 & 3 \end{array} \\ \begin{pmatrix} 0 & 1 & 1 \\ 1 & 1 & 0 \\ 1 & 1 & 0 \end{pmatrix} \end{array}.$$

A *trellis* $T = (V, E)$ *of width* N is a finite directed graph of a special kind. The *vertex set* V is the disjoint union of $N + 1$ finite sets $V_0, \ldots, V_N$:

$$V = V_0 \cup V_1 \cup \cdots \cup V_N.$$

The *edge set* E is the disjoint union of N finite sets $E_1, \ldots, E_N$, where $E_i \subseteq V_{i-1} \times V_i$:

$$E = E_1 \cup E_2 \cup \cdots \cup E_N.$$

Thus an edge $e \in E_i$ is of the form $e = (\alpha, \beta)$ with $\alpha \in V_{i-1}$, and $\beta \in V_i$. For example, Figure 2 shows a simple trellis with

$$V_0 = \{a, b, c\}, \ V_1 = \{d, e\}, \ V_2 = \{f, g, h, k\}$$

and

$$E_1 = \{(a, d), (b, d), (b, e), (c, e)\}, \ E_2 = \{(d, f), (d, g), (d, k), (e, f), (e, h), (e, k)\}.$$

Note that the trellis of Figure 2 also has edge weights.

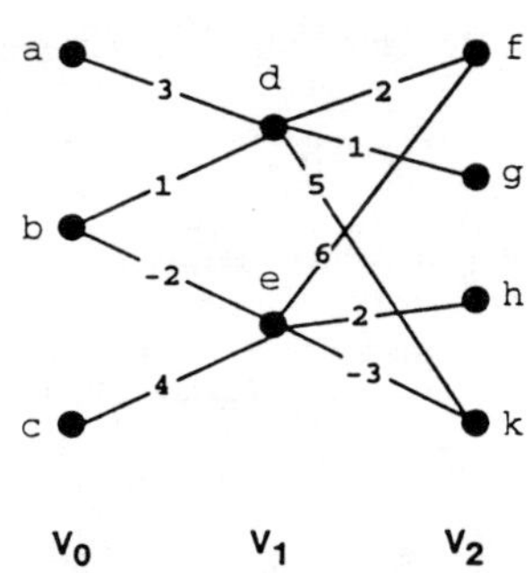

Figure 2. A simple weighted trellis of width 2.

A trellis T of width N with weighted edges can be represented by a sequence of N matrices $A_1, A_2, \ldots, A_N$. The ith matrix A_i has dimensions $|V_{i-1}| \times |V_i|$, with rows indexed by V_{i-1} and columns indexed by V_i. If $e = (\alpha, \beta) \in E_i$, with $\alpha \in V_{i-1}$ and $\beta \in V_i$, then the (α, β) entry of A_i is equal to $w(e)$. If there is no edge from α to β, the (α, β) entry of A_i is equal to 0. For example, the trellis of Figure 2 can be represented by the matrices A_1 and A_2, given by

$$A_1 = \begin{matrix} & d & e \\ a & 3 & 0 \\ b & 1 & -2 \\ c & 0 & 4 \end{matrix}, \quad A_2 = \begin{matrix} & f & g & h & k \\ d & 2 & 1 & 0 & 5 \\ e & 6 & 0 & 2 & -3 \end{matrix}. \tag{1.2}$$

We shall call $A_1, A_2, \ldots, A_N$ the *matrix representation* of the corresponding weighted trellis. Conversely, if $A_1, A_2, \ldots, A_N$ is a sequence of matrices over a semiring R, and if the matrices are compatible, i.e., if for each $1 \leq i < N$, the number of columns of A_i is the same as the number of rows of A_{i+1}, we can construct a weighted trellis of width N from the sequence $A_1, A_2, \ldots, A_N$ in a straightforward way. The trellis constructed this way is called the trellis *afforded* by the sequence $A_1, A_2, \ldots, A_N$.

If the matrices $A_1, A_2, \ldots, A_N$ are compatible in the sense just described, then the matrix product $A = A_1 A_2 \cdots A_N$ is defined. This suggests that the matrix product A may have combinatorial significance. This is indeed the case, as we shall now see.

If $u \in V_0$ and $v \in V_N$, we define the *flow* from u to v, denoted $\phi_T(u, v)$, to be the sum of the weights of all paths in T from u to v. The combinatorial quantity $\phi_T(u, v)$ is related to the matrix representation of T by the following simple result.

1.1 Theorem. *The flow $\phi_T(u, v)$ is equal to the (u, v)th entry of the product matrix*

$$A = A_1 A_2 \cdots A_N.$$

Proof: (See [1.Theorem 4.7.1.]) By definition of matrix multiplication.

$$A[i, j] = \sum A_1[i, i_1] A_2[i_1, i_2] \cdots A_N[i_{N-1}, j],$$

where the sum is over all possible sequences $(i_1, i_2, \ldots, i_{N-1})$ of indices in the matrices $A_1, \ldots, A_N$. This sum is zero unless there is a path of the form $(i, i_1, i_2, \ldots, i_{N-1}, j)$ from i to j in the trellis afforded by the sequence $A_1, \ldots, A_N$. If such a path exists, then the sum is equal to the sum of the weights of all such paths, which is what the theorem asserts.∎

For example, the product of the matrices A_1 and A_2 in (1.2) is

$$A = \begin{matrix} & f & g & h & k \\ a & 6 & 3 & 0 & 15 \\ b & -10 & 1 & -4 & 11 \\ c & 24 & 0 & 8 & -12 \end{matrix}.$$

As a partial check of Theorem 1.1 in this case, we note that in the trellis of Figure 2, there are no paths from a to h, so that $\phi_T(a,h) = 0$ which is also the (a,h) entry of A. Similarly, there are two paths from b to f, viz., (b,d,f) and (b,e,f), with path weights $1 \cdot 2 = 2$ and $-2 \cdot 6 = -12$ respectively. Thus $\phi_T(b,f) = -10$, which is indeed the (b,f) entry of A.

A common way to construct a trellis of width N is what we might call the *repeated concatenation* of N copies of a fixed weighted digraph. Thus let $D = (V,E)$ be a weighted digraph, and let B be its incidence matrix. Then the width N trellis constructed from D is the trellis afforded by the sequence $A_1, \ldots, A_N$, where $A_1 = A_2 = \cdots = A_N = B$. To distinguish between vertices at different depths in the trellis constructed this way, we introduce a "time index" t. Thus in the width-N trellis constructed from $D = (V,E)$, the vertex set is $V \times \{0,1,\ldots,N\}$, and the edge set E_i is

$$E_i = \{((\alpha, i-1), (\beta, i)) : (\alpha, \beta) \in E\}.$$

For example, if we start with the digraph in Figure 1, whose incidence matrix is given by (1.1), we can construct the width-4 trellis shown in Figure 3, where the "time index" is shown along the bottom.

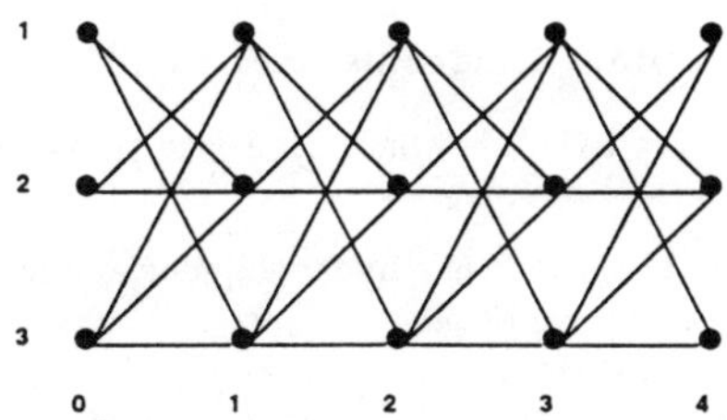

Figure 3. The width-4 trellis constructed from the digraph of Figure 1.

We conclude this section with what Stanley [1] calls the *transfer-function* theorem for weighted digraphs. Let D be a weighted digraph with incidence matrix A, and for $N \geq 0$ let $A_{ij}(N)$ denote the sum of the weights of all paths in D of length N from v_i to v_j.[2] The quantities $A_{ij}(N)$ can be viewed as the entries of an $M \times M$ matrix $A(N)$. If we form the matrix generating function

$$F_G(z) = \sum_{N \geq 0} A(N) z^N, \tag{1.3}$$

we have the following basic result.

[2] We define $A_{ij}(0)$ to be 1 if $i = j$ and 0 if $i \neq j$.

1.2 Theorem. *(The transfer-function theorem) The generating function $F_G(z)$ is given by*

$$F_G(z) = (I - zA)^{-1}, \tag{1.4}$$

where in (1.4), I denotes the $M \times M$ identity matrix.

Proof: First note that the quantity $A_{ij}(N)$ is the sum of all paths from $(v_i, 0)$ to (v_j, N) in the trellis of width N constructed from D. Thus by Theorem 1.1, $A_{ij}(N)$ is the (i, j)th entry in the matrix A^N, i.e., $A(N) = A^N$. Thus

$$\begin{aligned} F_G(z) &= \sum_{N \geq 0} A(N) z^N \\ &= \sum_{N \geq 0} A^N z^N \\ &= (I - zA)^{-1}, \end{aligned}$$

using the "sum of a geometric series" theorem for matrices. ▪

2. Convolutional Codes, State Diagrams, and Incidence Matrices.

In this section we briefly review the state-space approach to convolutional codes, and state the counting problems we intend to solve.

An (n, k, m) *convolutional encoder* is a linear sequential device which maps a sequence of k-dimensional *information*, or *input blocks* $\boldsymbol{\alpha}_0, \boldsymbol{\alpha}_1, \ldots$ over the finite field $GF(2)$ into a sequence of n-dimensional *output*, or *code blocks* $\boldsymbol{\beta}_0, \boldsymbol{\beta}_1, \ldots$. The encoder also passes through a sequence of internal m-dimensional *state vectors*, or simply *states* $\boldsymbol{\sigma}_0, \boldsymbol{\sigma}_1 \ldots$. The jth code block $\boldsymbol{\beta}_j$ is a linear function of the jth input block $\boldsymbol{\alpha}_j$, and also of the jth state $\boldsymbol{\sigma}_j$. The formal description of the encoder's operation is this: the initial state $\boldsymbol{\sigma}_0$ is arbitrary and for $j \geq 0$,

$$\boldsymbol{\sigma}_{j+1} = \boldsymbol{\sigma}_j \mathcal{A} + \boldsymbol{\alpha}_j \mathcal{B} \tag{2.1}$$

$$\boldsymbol{\beta}_j = \boldsymbol{\sigma}_j \mathcal{C} + \boldsymbol{\alpha}_j \mathcal{D}, \tag{2.2}$$

where the four matrices $\mathcal{A}$, $\mathcal{B}$, $\mathcal{C}$, and $\mathcal{D}$ have entries from $GF(2)$ and dimensions

$$\begin{aligned} \mathcal{A} &: m \times m \\ \mathcal{B} &: k \times m \\ \mathcal{C} &: m \times n \\ \mathcal{D} &: k \times n. \end{aligned} \tag{2.3}$$

The (n, k, m) *convolutional code* associated with such an encoder is then defined to be the set of all possible output sequences (or *codewords*) $(\boldsymbol{\beta}_0, \boldsymbol{\beta}_1, \ldots)$ which can be produced form the encoder, starting with the initial state $\boldsymbol{\sigma}_0 = \mathbf{0}$.

For example, consider the $(2, 1, 2)$ encoder shown in Figure 4. This encoder has input α, output $\boldsymbol{\beta} = (\beta_1, \beta_2)$ and state $\boldsymbol{\sigma} = (\sigma_1, \sigma_2)$. The next state will be $\boldsymbol{\sigma}' = (\alpha+\sigma_1+\sigma_2, \sigma_1)$, and the output $\boldsymbol{\beta}$ is given by $(\beta_1, \beta_2) = (\alpha, \alpha + \sigma_1)$, so the four matrices $\mathcal{A}$, $\mathcal{B}$, $\mathcal{C}$, and $\mathcal{D}$ are, in this case,

$$\mathcal{A} = \begin{pmatrix} 1 & 1 \\ 1 & 0 \end{pmatrix}, \quad \mathcal{B} = (1 \quad 0)$$

$$\mathcal{C} = \begin{pmatrix} 0 & 1 \\ 0 & 0 \end{pmatrix}, \quad \mathcal{D} = (1 \quad 1).$$

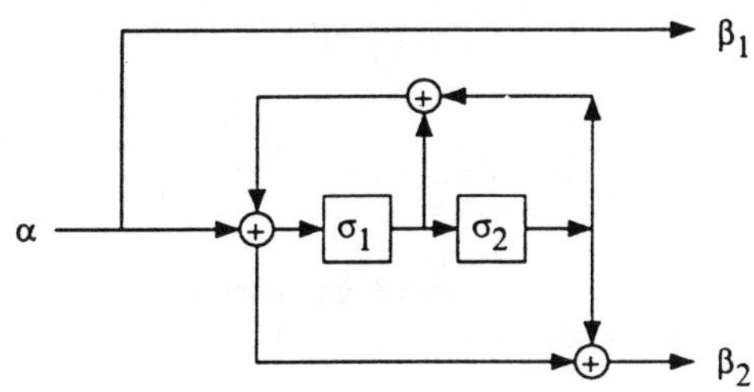

Figure 4. A simple $(2, 1, 2)$ convolutional encoder.

Associated with each convolutional encoder is a corresponding *state diagram*, which is a weighted digraph of the type discussed in Section 1. The state diagram has 2^m vertices, one for each of the 2^m states of the encoder. For each state $\boldsymbol{\sigma}$ and input block $\boldsymbol{\alpha}$, there is a doubly-labelled edge from $\boldsymbol{\sigma}$ to $\boldsymbol{\sigma}'$:

(2.4)
$$\boldsymbol{\sigma} \xrightarrow[\boldsymbol{\beta}]{\boldsymbol{\alpha}} \boldsymbol{\sigma}'$$

where

$$\boldsymbol{\sigma}' = \boldsymbol{\sigma}\mathcal{A} + \boldsymbol{\alpha}\mathcal{B}$$
$$\boldsymbol{\beta} = \boldsymbol{\sigma}\mathcal{C} + \boldsymbol{\alpha}\mathcal{D}.$$

Thus the state diagram has 2^m vertices and 2^{m+k} edges. Figure 5 shows the state diagram for the encoder shown in Figure 4.

A semi-infinite *codepath* in the state diagram is a sequence of edges in the state diagram, say $e_1 e_2 \cdots$, where for each index $i = 1, 2, \ldots$, $\text{fin}(e_i) = \text{init}(e_{i+1})$. Such a codepath can be represented as follows:

(2.5)
$$\boldsymbol{\sigma}_0 \xrightarrow[\boldsymbol{\beta}_0]{\boldsymbol{\alpha}_0} \boldsymbol{\sigma}_1 \xrightarrow[\boldsymbol{\beta}_1]{\boldsymbol{\alpha}_1} \boldsymbol{\sigma}_2 \xrightarrow[\boldsymbol{\beta}_2]{\boldsymbol{\alpha}_2} \cdots$$

The codepath in (2.5) is said to have *initial state* $\boldsymbol{\sigma}_0$ and to be *generated by the input sequence* $(\boldsymbol{\alpha}_0, \boldsymbol{\alpha}_1, \ldots)$. The path is said to *produce the output sequence* $(\boldsymbol{\alpha}_0, \boldsymbol{\alpha}_1, \ldots)$. The

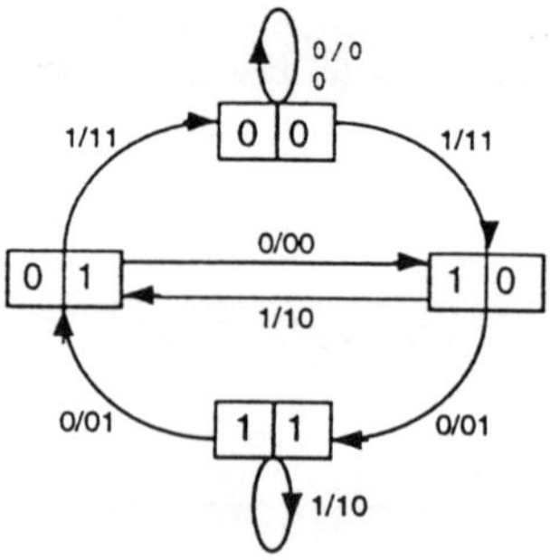

Figure 5. The state diagram for the encoder of Figure 4.

convolutional code associated with the encoding matrices $(\mathcal{A}, \mathcal{B}, \mathcal{C}, \mathcal{D})$ is thus the set of all possible output sequences produced by semi-infinite codepaths of the form (2.5) which have initial state $\boldsymbol{\sigma}_0 = \mathbf{0}$.

A *finite codepath of length N* is a codepath of the form

$$\boldsymbol{\sigma}_0 \xrightarrow[\boldsymbol{\beta}_0]{\boldsymbol{\alpha}_0} \boldsymbol{\sigma}_1 \xrightarrow[\boldsymbol{\beta}_1]{\boldsymbol{\alpha}_1} \boldsymbol{\sigma}_2 \xrightarrow[\boldsymbol{\beta}_2]{\boldsymbol{\alpha}_2} \cdots \boldsymbol{\sigma}_{N-1} \xrightarrow[\boldsymbol{\beta}_{N-1}]{\boldsymbol{\alpha}_{N-1}} \boldsymbol{\sigma}_N. \tag{2.6}$$

In this paper we shall count finite codepaths of several types. Here are the four basic types we shall consider.

- The *closed codepaths*, i.e., the codepaths whose initial and final states are the same.
- The 0-*closed codepaths*, i.e., the codepaths whose initial and final states are both equal to $\mathbf{0}$.
- The *molecular codepaths*, i.e., the 0-closed codepaths which do not use the *zero edge* $\mathbf{0} \xrightarrow[\mathbf{0}]{\mathbf{0}} \mathbf{0}$.
- The *atomic codepaths*, i.e., the molecular codepaths which do not enter the zero state, except at the beginning and end.

We shall classify these codepaths according to two, and sometimes three, numerical attributes: the *length*, the *output weight*, and (sometimes) the *input weight*.[3] We shall do this classification using the method of generating functions, which are sometimes called

[3] It is unfortunate that the term *weight* has two different meanings in this paper. The weight of a path in a digraph is of course a very different thing from the weight of a vector or codeword. Both uses are well-established, however, and we have chosen to risk ambiguity rather than introduce new terminology.

weight enumerators in this context. For example, we if we denote by $w_{N,i,j}$ the number of 0-closed paths of length N, input weight i and output weight j, we will seek a closed-form expression for the three-variable generating function

$$W(x,y,z) = \sum_{N,i,j} w_{N,i,j} x^i y^j z^N.$$

The basic tools we use to count these codepaths are the transfer matrix theorem method described in Section 1 and xy-labelled and the y-labelled state diagrams of the encoder, and the corresponding incidence matrices.

The xy-weight state diagram of the code (also called the xy-state diagram) is the diagram obtained by replacing the label $(\boldsymbol{\alpha}, \boldsymbol{\beta})$ on each edge of the state diagram (see (2.4)) with the label $x^{|\alpha|}y^{|\beta|}$, where $|\boldsymbol{x}|$ denotes the weight of the vector $\boldsymbol{x}$. Thus the exponent of x represents the input weight, and the exponent of y represents the output weight, of the corresponding state transition, so that the weight of a codepath in the xy-state diagram is of the form $x^{w_I}y^{w_O}$, where w_I is the input weight, and w_O is the output weight, of the path. Similarly, the y-weight state diagram (also called the y-state diagram) is obtained by replacing the $(\boldsymbol{\alpha}, \boldsymbol{\beta})$ label with $y^{|\beta|}$, so that the weight of a path in the y-state diagram is simply y^{w_O}. In Figures 6 and 7, we see the xy-weight and y-weight state diagrams for the encoder of Figure 4.

For calculation purposes, the xy- and y-weight state diagrams can be be represented by the corresponding incidence matrices. For example, the xy-incidence matrix corresponding to the xy-state diagram of Figure 6 is

$$(2.7) \qquad A(x,y) = \begin{array}{c} \\ 00 \\ 10 \\ 01 \\ 11 \end{array} \begin{array}{c} \begin{array}{cccc} 00 & 10 & 01 & 11 \end{array} \\ \begin{pmatrix} 1 & xy^2 & 0 & 0 \\ 0 & 0 & xy & y \\ xy^2 & 1 & 0 & 0 \\ 0 & 0 & y & xy \end{pmatrix} \end{array},$$

and the y-incidence matrix corresponding to the y-state diagram of Figure 7 is

$$(2.8) \qquad A(y) = \begin{array}{c} \\ 00 \\ 10 \\ 01 \\ 11 \end{array} \begin{array}{c} \begin{array}{cccc} 00 & 10 & 01 & 11 \end{array} \\ \begin{pmatrix} 1 & y^2 & 0 & 0 \\ 0 & 0 & y & y \\ y^2 & 1 & 0 & 0 \\ 0 & 0 & y & y \end{pmatrix} \end{array}.$$

Finally, in our study of the molecular and atomic codepaths, we will need the *expurgated* xy incidence matrix for the code, denoted $A^\bullet(x,y)$. This matrix is identical to its counterpart $A(x,y)$, except that the $(0,0)$ entry is reduced by 1, which is the contribution to $A(x,y)$ of the *zero edge*

$$0 \xrightarrow[0]{0} 0.$$

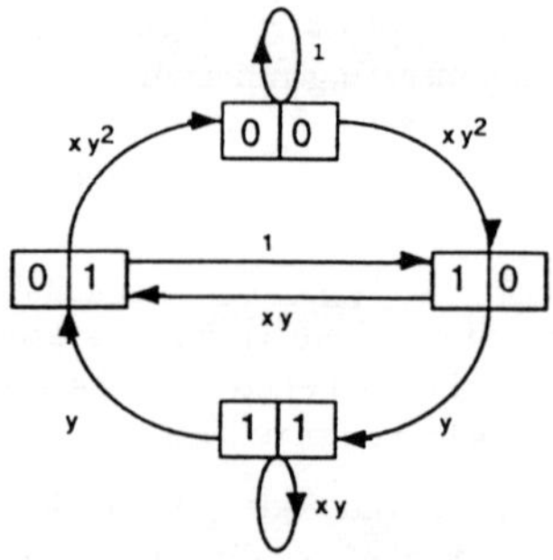

Figure 6. The xy-state diagram for the encoder of Figure 4.

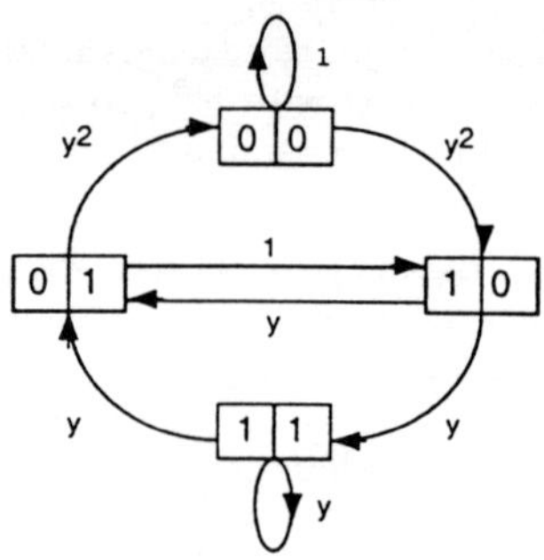

Figure 7. The y-state diagram for the encoder of Figure 4.

For example, for the encoder of Figure 4,

$$A^\bullet(x,y) = \begin{array}{c} \\ 00 \\ 10 \\ 01 \\ 11 \end{array} \begin{array}{c} \begin{array}{cccc} 00 & 10 & 01 & 11 \end{array} \\ \begin{pmatrix} 0 & xy^2 & 0 & 0 \\ 0 & 0 & xy & y \\ xy^2 & 1 & 0 & 0 \\ 0 & 0 & y & xy \end{pmatrix} \end{array} . \tag{2.9}$$

3. Molecular and Atomic Paths. The Transfer Function.

In this section we will find the xy-weight enumerators for the set of molecular and atomic codepaths. Specifically, we will investigate the generating functions

$$V^\bullet(x,y,z) = \sum_{N\geq 0} V_N^\bullet(x,y)z^N \tag{3.1}$$

$$W^\bullet(x,y,z) = \sum_{N\geq 0} W_N^\bullet(x,y)z^N, \tag{3.2}$$

where in (3.1), $V_N^\bullet(x,y)$ denotes the xy weight enumerator for the molecular codepaths of length N, and $W_N^\bullet(x,y)$ denotes the xy weight enumerator for the atomic codepaths of length N. The trivariate generating function $W^\bullet(x,y,z)$ is often called the *transfer function* for the code, and is useful for estimating the code's performance on a memoryless channel [2, Sec. 4.3].

3.1 Theorem. *We have*

$$V^\bullet(x,y,z) = \left[(I - zA^\bullet(x,y))^{-1}\right]_{0,0} \tag{3.3}$$

$$W^\bullet(x,y,z) = 1 - \frac{1}{[(I - zA^\bullet(x,y))^{-1}]_{0,0}}. \tag{3.4}$$

Proof: Since the molecular codepaths are, by definition, the 0-closed codepaths which do not use the zero edge, the trellis for the xy-weight enumerator for the molecular codepaths of length N is the one afforded by the matrix sequence

$$(A^\bullet(x,y),\ldots,A^\bullet(x,y)) \qquad (N \text{ terms}),$$

it follows from the Theorem 1.1 that the weight enumerator $V_N^\bullet(x,y)$ is given by the formula

$$V_N^\bullet(x,y) = \left[A^\bullet(x,y)^N\right]_{0,0}.$$

Thus we have

$$\begin{aligned} V^\bullet(x,y,z) &= \sum_{N\geq 0} V_N^\bullet(x,y)z^N \\ &= \sum_{N\geq 0} \left[A^\bullet(x,y)^N\right]_{0,0} z^N \\ &= \left[\sum_{N\geq 0} A^\bullet(x,y)^N z^N\right]_{0,0} \\ &= \left[(I - zA^\bullet(x,y))^{-1}\right]_{0,0}, \end{aligned}$$

which proves (3.3).

To prove (3.4), note that each molecular codepath can be "factored" uniquely into a concatenation of a finite number (possibly zero) of atomic codepaths. Thus, since $W^\bullet(x,y,z)$ denotes the enumerator for the atomic codepaths, we have

$$\begin{aligned} V^\bullet(x,y,z) &= 1 + W^\bullet(x,y,z) + W^\bullet(x,y,z)^2 + W^\bullet(x,y,z)^3 + \cdots \\ &= 1/(1 - W^\bullet(x,y,z)). \end{aligned}$$

Thus $W^\bullet(x,y,z) = 1 - 1/V^\bullet(x,y,z)$, which, given (3.3), proves (3.4). ▪

We conclude this section by illustrating the results for the encoder of Figure 4. Here $A^\bullet(x,y)$ is given by (2.9). Using *Mathematica*, or a similar symbolic manipulation program, it is then easy to apply (3.3) to find that

$$V^\bullet(x,y,z) = \frac{1 - xyz - xyz^2 + (-y^2 + x^2y^2)z^3}{1 - xyz - xyz^2 + (-y^2 + x^2y^2 - x^3y^5)z^3 + (-x^2y^6 + x^4y^6)z^4},$$

and

$$W^\bullet(x,y,z) = \frac{x^3y^5z^3 + (x^2y^6 - x^4y^6)z^4}{1 - xyz - xyz^2 + (-y^2 + x^2y^2)z^3}. \tag{3.5}$$

It follows then from the form of (3.5), that the individual weight enumerators $W_N^\bullet(x,y)$ in (3.2) satisfy the recursion

$$W_N^\bullet(x,y) = xyW_{N-1}^\bullet(x,y) + xyW_{N-2}^\bullet(x,y) + (y^2 - x^2y^2)W_{N-3}^\bullet(x,y) \qquad \text{for } N \geq 6. \tag{3.6}$$

We can obtain the first few terms of the expansion of $W^\bullet(x,y,z)$ in powers of z using long division:

$$W^\bullet(x,y,z) = x^3y^5z^3 + x^2y^6z^4 + (x^4y^6 + x^3y^7)z^5 + (\text{terms of order } z^6 \text{ and higher}),$$

which gives the entries in rows 3, 4, and 5 in Table 1. Given these three rows as initial conditions, using the recursion (3.6), we can obtain (by computer) as many rows in Table 1 as desired. (In Table 1, we have represented the function $W_N^\bullet(x,y)$ as $\sum_j w_{N,j}^\bullet(x)y^j$, where $w_{N,j}^\bullet$ is a polynomial in x.)

$N\backslash j$	5	6	7	8	9	10	11
0							
1							
2							
3	x^3						
4		x^2					
5		x^4	x^3				
6			$2x^3$	x^4			
7			x^5	$x^2 + 2x^4$	x^5		
8				$3x^4$	$2x^3 + 2x^5$	x^6	
9				x^6	$3x^3 + 3x^5$	$3x^4 + 2x^6$	x^7
$\vdots$							

Table 1. The polynomials $w^{\bullet}_{N,j}(x)$ for the encoder of Figure 4.

4. Truncated and Tailbiting Codes, Output Weight Only.

In this section we will find generating functions for the set of 0-closed and closed codepaths, which will give us the weight enumerators for the truncated and tailbiting[4] block codes derived from the parent convolutional code. Specifically, we will investigate the generating functions

$$W(y,z) = \sum_{N\geq 0} W_N(y)z^N \tag{4.1}$$

$$W^{\bullet}(y,z) = \sum_{N\geq 0} W^{\bullet}_N(y)z^N, \tag{4.2}$$

where in (4.1), $W_N(y)$ denotes the y weight enumerator for the 0-closed codepaths of length N, and $W^{\bullet}_N(y)$ denotes the y weight enumerator for the set of closed codepaths of length N.

4.1 Theorem. *We have*

$$W(y,z) = \left[(I - zA(y))^{-1}\right]_{0,0} \tag{4.3}$$

$$W^{\bullet}(y,z) = \operatorname{Tr}\left[(I - zA(y))^{-1}\right]. \tag{4.4}$$

[4] The Nth *truncated code* derived from a parent convolutional code is the set of codewords generated by all finite codepaths of length N whose initial and final states are both 0. The Nth *tailbiting* code is the set of codewords generated by all finite codepaths of length N whose initial and final states are the same.

Proof: Since the trellis for the y-weight enumerator for the closed codepaths of length N is the one afforded by the matrix sequence

$$(A(y), \ldots, A(y)) \qquad (N \text{ terms}),$$

it follows from Theorem 1.1 that the weight enumerator $W_N(y)$ is given by

$$W_N(y) = \left[A(y)^N\right]_{0,0}.$$

Thus we have

$$\begin{aligned} W(y,z) &= \sum_{N\geq 0} W_N(y) z^N \\ &= \sum_{N\geq 0} \left[A(y)^N\right]_{0,0} z^N \\ &= \left[\sum_{N\geq 0} A(y)^N z^N\right]_{0,0} \\ &= \left[(I - zA(y))^{-1}\right]_{0,0}, \end{aligned}$$

which proves (4.3). The result (4.4) is proved similarly.[5] ∎

The following result gives an elegant alternative method for calculating the enumerators $W_N^\bullet(y)$.

4.2 Corollary. *We have*

$$\sum_{N\geq 1} W_N^\bullet(y) z^N = -\frac{z\frac{\partial Q(y,z)}{\partial z}}{Q(y,z)}, \tag{4.5}$$

where $Q(y,z) = \det(I - zA(y))$. (Note that the sum in (4.5) begins at $N = 1$, whereas the sum in (4.2) begins at $N = 0$.)

Proof: This follows immediately from Corollary 4.7.3 in [1], which proves that in an arbitrary labelled digraph with incidence matrix A,

$$\sum_{N\geq 1} C_G(N) z^N = -\frac{zQ'(z)}{Q(z)},$$

where $C_G(N)$ denotes the sum of the weights of all closed paths of length N and $Q(z) = \det(I - zA)$. ∎

[5] In (4.4), the notation "Tr" represents the trace of a square matrix, i.e., the sum of the diagonal entries.

We conclude this section by illustrating the results for the encoder of Figure 4. Here $A(y)$ is given by (2.8). Using *Mathematica* and Theorem 4.1, we find that

$$W(y, z) = \frac{1 - yz - yz^2}{1 - (1 + y)z - (-y + y^5)z^3}. \tag{4.6}$$

From (4.1) and (4.6), we see that the individual weight enumerators $W_N(y)$ satisfy the recursion

$$W_N(y) = (1 + y)W_{N-1}(y) + (-y + y^5)W_{N-3}(y). \qquad \text{for } N \geq 3. \tag{4.7}$$

Expanding (4.6) in ascending powers of z, either by hand or by using *Mathematica*, we find

$$W(y, z) = 1 + z + z^2 + (\text{terms of order } z^3 \text{ and higher})$$

which gives the initial conditions

$$W_0(y) = W_1(y) = W_2(y) = 1,$$

which appear as the first three rows of Table 2. The remainder of Table 2 is then easily found using the recursion (4.7). (In Table 2, we have represented the function $W_N(y)$ as $\sum_j w_{N,j}y^j$.)

$N\backslash j$	0	1	2	3	4	5	6	7	8	9	10	11	12	13	14	15	⋯
0	1																
1	1																
2	1																
3	1					1											
4	1					2	1										
5	1					3	3	1									
6	1					4	5	4	1		1						
7	1					5	7	8	5	1	3	2					
8	1					6	9	12	12	6	7	8	3				
9	1					7	11	16	20	17	17	19	15	4		1	
⋮																	

Table 2. The numbers $w_{N,j}$ for the encoder of Figure 4.

Similarly, an application of Theorem 4.1 shows that

$$W^*(y, z) = \frac{4 - (3 + 3y)z - (-y + y^5)z^3}{1 - (1 + y)z - (-y + y^5)z^3}. \tag{4.8}$$

From (4.2) and (4.8), we see that the individual weight enumerators $W_N^*(y)$ satisfy the same recursion as the $W_N(y)$'s, viz.,

$$W_N^*(y) = (1+y)W_{N-1}^*(y) + (-y+y^5)W_{N-3}^*(y), \qquad \text{for } N \geq 3. \tag{4.9}$$

Expanding (4.8) in ascending powers of z, either by hand or by using *Mathematica*, we find

$$W^*(y,z) = 4 + (1+y)z + (1+2y+y^2)z^2 + (\text{terms of order } z^3 \text{ and higher}),$$

i.e.,

$$\begin{aligned} W_0^*(y) &= 4 \\ W_1^*(y) &= 1+y \\ W_2^*(y) &= 1+2y+y^2 \end{aligned}$$

which gives the first three rows of Table 3. It is then easy to complete Table 3 using the recursion (4.9). (In Table 3, we have represented the function $W_N^*(y)$ as $\sum_j w_{N,j}^* y^j$.)

$N\backslash j$	0	1	2	3	4	5	6	7	8	9	10	11	12	13	14	15	$\cdots$
0	4																
1	1	1															
2	1	2	1														
3	1		3	1		3											
4	1		2	4	1	4	4										
5	1			5	5	6	10	5									
6	1			2	9	12	13	18	6		3						
7	1				7	21	21	29	28	7	7	7					
8	1				2	24	36	40	57	40	20	24	12				
9	1					18	48	63	81	100	72	54	54	18		3	
$\vdots$																	

Table 3. The numbers $w_{N,j}^*$ for the encoder of Figure 4.

If we use the alternative method of computing the $W_N^*(y)$'s given in Corollary 4.2, we find that the polynomial $Q(y,z) = \det(I - zA(y))$, where $A(y)$ is given by (2.8), is

$$Q(y,z) = 1 - (1+y)z - (-y+y^5)z^3. \tag{4.10}$$

Thus from Corollary 4.2, we have

$$\sum_{N \geq 1} W_N^*(y) z^N = -\frac{z\frac{\partial Q(y,z)}{\partial z}}{Q(y,z)} = \frac{(1+y)z + (-3y+3y^5)z^3}{1-(1+y)z-(-y+y^5)z^3}. \tag{4.11}$$

From (4.11), we again see that the recursion (4.9) holds, and so are led to the expansion

$$\sum_{N\geq 1} W_N^*(y)z^N = (1+y)z + (1+2y+y^2)x^2 + (1+3y^2+y^3+3y^5)z^3 + \cdots,$$

which gives (except for $N = 0$) the same values for the $W_N^*(y)$'s as the method of Theorem 4.1.

5. Truncated and Tailbiting Codes, Input-Output Weight Enumerators.

In this section we restrict ourselves to convolutional codes with $k = 1$. We will find xyz-generating functions for the set of 0-closed and closed codepaths, which will give us the input-output weight enumerators for the truncated and tailbiting block codes derived from the parent convolutional code (with $k = 1$). Specifically, we will investigate the generating functions

$$W(x,y,z) = \sum_{N\geq m} W_N(x,y)z^N \tag{5.1}$$

$$W^*(x,y,z) = \sum_{N\geq m} W_N^*(x,y)z^N, \tag{5.2}$$

where in (5.1), $W_N(x,y)$ denotes the xy weight enumerator for the set of 0-closed codepaths of length N, and $W_N^*(x,y)$ denotes the xy weight enumerator for the set of closed codepaths of length N. We assume that the encoder's input is "turned off" after $N-m$ clock cycles and that the encoder is then forced into the appropriate terminal state during the final m clock cycles.

5.1 Theorem. *We have*

$$W(x,y,z) = z^m\left\{(I - zA(x,y))^{-1}A(y)^m\right\}_{(0,0)} \tag{5.3}$$

$$W^*(x,y,z) = z^m \operatorname{Tr}\left\{\Delta(x)(I - zA(x,y))^{-1}A^m(y)\right\}. \tag{5.4}$$

where in (5.4), $\Delta(x)$ is a diagonal matrix whose (s,s)th entry is $x^{|s|}$.

Proof: According to our assumption, for a truncated or a tailbiting code of length N, the encoder accepts inputs for the first $N-m$ times clock cycles, at which point the input is turned off for the remaining m clock cycles so that the trellis path can be properly terminated. Thus for computing $W_N(x,y)$ and $W_N^*(x,y)$, the appropriate trellis is the one afforded by the matrix sequence

$$A(x,y), A(x,y), \ldots, A(x,y), A(y), \ldots, A(y),$$

where there are $N-m$ terms equal to $A(x,y)$, and m terms equal to $A(y)$. It follows then from Theorem 1.1 that

$$W_N(x,y) = \left[A(x,y)^{N-m}A(y)^m\right]_{0,0}.$$

Thus we have

$$
\begin{aligned}
W(x,y,z) &= \sum_{N\geq m} W_N(x,y)z^N \\
&= \sum_{N\geq m} \left[A(x,y)^{N-m}A(y)^m\right]_{0,0} z^N \\
&= z^m \left[\sum_{N\geq m} z^{N-m}A(x,y)^{N-m}A(y)^m\right]_{0,0} \\
&= z^m \left[\left(\sum_{h\geq 0} A(x,y)^h z^h\right) A(y)^m\right]_{0,0} \\
&= z^m \left\{(I - zA(x,y))^{-1}A(y)^m\right\}_{(0,0)},
\end{aligned}
$$

which proves (5.3).

To prove (5.3), we note that the first m information bits in the encoder for a tailbiting code of length N are used to select the initial state, at which point the encoder behaves like the encoder for the truncated code. Thus

$$
W^*_N(x,y) = \sum_s x^{|s|} \left[A(x,y)^{N-m}A(y)^m\right]_{0,0}.
$$

Thus we have

$$
\begin{aligned}
W^*(x,y,z) &= \sum_{N\geq m} W^*_N(x,y)z^N \\
&= \sum_{N\geq m} z^N \sum_s x^{|s|} \left[A(x,y)^{N-m}A(y)^m\right]_{0,0} \\
&= \sum_{N\geq m} z^N \operatorname{Tr}\left\{\Delta(x)A(x,y)^{N-m}A(y)^m\right\} \\
&= \operatorname{Tr}\left\{ z^m\Delta(x)\left(\sum_{h\geq 0} A(x,y)^h z^h\right) A(y)^m\right\} \\
&= z^m \operatorname{Tr}\left\{\Delta(x)[I - zA(x,y)]^{-1}A(y)^m\right\}.
\end{aligned}
$$

which proves (5.3).▪

We conclude this section by illustrating the results for the encoder of Figure 4. Here $A(x,y)$ is given by (2.7), and $A(y)$ is given by (2.8). Using *Mathematica*, it is then easy to apply (5.3) to find that

$$
W(x,y,z) = \frac{P(x,y,z)}{Q(x,y,z)},
$$

where

$$P(x,y,z) = z^2 + (-xy + xy^5)z^3 + (-xy + x^2y^5 + xy^6 - x^2y^6)z^4 \\ + (-y^2 + x^2y^2 + xy^6 - x^3y^6)z^5 \tag{5.5}$$

and

$$Q(x,y,z) = 1 + (-1 - xy)z + (xy - y^2 + x^2y^2 - x^3y^5)z^3 \\ + (y^2 - x^2y^2 - x^2y^6 + x^4y^6)z^4. \tag{5.6}$$

Using *Mathematica*, we find that the first few terms in the expansion of $W(x,y,z)$ in powers of z are

$$\begin{aligned} W(x,y,z) = z^2 &+ (1 + xy^5)z^3 + (1 + (x + x^2)y^5 + xy^6)z^4 \\ &+ (1 + (x + x^2 + x^3)y^5 + (2x + x^2)y^6 + x^2y^7)z^5 \\ &+ \text{terms of order } z^6 \text{ and higher} \end{aligned} \tag{5.7}$$

The remaining terms in the expansion of $W(x,y,z)$ can be obtained by noting that the form of the denominator in (5.6) implies that the individual xy-weight enumerators $W_N = W_N(x,y)$ satisfy the fourth-order recursion

$$\begin{aligned} W_N = (1 + xy)W_{N-1} &+ (-xy + y^2 - x^2y^2 + x^3y^5)W_{N-3} \\ &+ (-y^2 + x^2y^2 + x^2y^6 - x^4y^6)W_{N-4} \qquad \text{for } N \geq 6. \end{aligned} \tag{5.8}$$

Hence using the initial conditions provided by (5.7), viz.,

$$\begin{aligned} W_2(x,y) &= 1 \\ W_3(x,y) &= 1 + xy^5 \\ W_4(x,y) &= 1 + (x + x^2)y^5 + xy^6 \\ W_5(x,y) &= 1 + (x + x^2 + x^3)y^5 + (2x + x^2)y^6 + x^2y^7 \end{aligned}$$

and the recursion (5.8), we obtain the values shown in Table 4, which may easily be extended as far as desired. (In Table 4, we have represented the function $W_N(x,y)$ as $\sum_j w_{N,j}(x)y^j$, where $w_{N,j}$ is a polynomial in x.)

$N\backslash j$	0	5	6	7	8	9	10	11
2	1							
3	1	x						
4	1	$x + x^2$	x					
5	1	$x + x^2 + x^3$	$2x + x^2$	x^2				
6	1	$x + x^2 + 2x^3$	$2x + 2x^2 + x^3$	$x + 3x^2$	x^3		x^4	
7	1	$x + x^2 + 3x^3$	$2x + 3x^2 + x^3 + x^4$	$x + 5x^2 + 2x^3$	$x + x^2 + 3x^3$	x^4	$2x^4 + x^5$	$x^3 + x^4$
⋮								

Table 4. The polynomials $w_{N,j}(x)$ for the encoder of Figure 4.

Similarly, using (5.4), we find

$$W^*(x,y,z) = \frac{P^*(x,y,z)}{Q^*(x,y,z)}, \tag{5.9}$$

where

$$\begin{aligned}(5.10)\,P^*(x,y,z) &= (1+2xy+x^2y^2)z^2 \\ &+ (-3xy+2xy^2-2x^2y^2+xy^5+2x^2y^5)\, z^3 \\ &+ (-xy-2xy^2+x^2y^2+xy^3-x^3y^3+x^3y^5+2xy^6-x^2y^6)\, z^4 \\ &+ (-y^2+x^2y^2-xy^3+x^3y^3+x^2y^6-x^4y^6+x^3y^7-x^5y^7)\, z^5\end{aligned}$$

and

$$\begin{aligned}(5.11)\; Q^*(x,y,z) &= 1+(-1-xy)\, z+(xy-y^2+x^2y^2-x^3y^5)\, z^3 \\ &+ (y^2-x^2y^2-x^2y^6+x^4y^6)\, z^4.\end{aligned}$$

Using *Mathematica*, we find that the first few terms in the expansion of $W^*(x,y,z)$ in powers of z are

$$\begin{aligned}W^*(x,y,z) &= (1+2xy+x^2y^2)z^2+(1+2xy^2+x^2y^2+x^3y^3+xy^5+2x^2y^5)z^3 \\ &+ (1+2x^2y^2+xy^3+2x^2y^3+x^3y^3+x^4y^4 \\ &\quad +xy^5+2x^2y^5+x^3y^5+2xy^6+2x^3y^6)z^4 \\ &+ (1+2xy^3+2x^2y^3+x^3y^3+2x^2y^4+2x^3y^4 \\ &\quad +x^4y^4+xy^5+2x^2y^5+2x^3y^5+x^5y^5+2xy^6+2x^2y^6 \\ &\quad +4x^3y^6+2x^4y^6+2x^2y^7+x^3y^7+2x^4y^7)z^5 \\ &+ \text{terms of order } z^6 \text{ and higher}\end{aligned} \tag{5.12}$$

The remaining terms in the expansion of $W^*(x,y,z)$ can be obtained by noting that since the denominator $Q^*(x,y,z)$ is the same as $Q(x,y,z)$ (cf. (5.6) and (5.11)), the individual xy-weight enumerators $W_N^* = W_N^*(x,y)$ satisfy the same fourth-order recursion as $W_N(x,y)$:

$$\begin{aligned}W_N^* &= (1+xy)W_{N-1}^* + (-xy+y^2-x^2y^2+x^3y^5)W_{N-3}^* \\ &+ (-y^2+x^2y^2+x^2y^6-x^4y^6)W_{N-4}^* \qquad \text{for } N \geq 6.\end{aligned} \tag{5.13}$$

Hence using the initial conditions provided by (5.7), viz.,

$$\begin{aligned}W_2^*(x,y) &= 1+2xy+x^2y^2 \\ W_3^*(x,y) &= 1+2xy^2+x^2y^2+x^3y^3+xy^5+2x^2y^5 \\ W_4^*(x,y) &= 1+2x^2y^2+xy^3+2x^2y^3+x^3y^3+x^4y^4+xy^5+2x^2y^5+x^3y^5+2xy^6+2x^3y^6 \\ W_5^*(x,y) &= 1+2xy^3+2x^2y^3+x^3y^3+2x^2y^4+2x^3y^4+x^4y^4+xy^5+2x^2y^5+2x^3y^5 \\ &\quad +x^5y^5+2xy^6+2x^2y^6+4x^3y^6+2x^4y^6+2x^2y^7+x^3y^7+2x^4y^7.\end{aligned}$$

and the recursion (5.13), we can extend this series as far as desired. (We omit any further terms, however, since e.g. $W_6^*(x,y)$ has 26 terms.)

References.

1. R. P. Stanley, *Enumerative Combinatorics, vol. I.* Monterey, Calif.: Wadsworth & Brooks/Cole, 1986.

2. A. J. Viterbi and J. K. Omura, *Principles of Digital Communication and Coding.* New York: McGraw-Hill, 1979.

11. Convolutional Codes and Complexity

A.J. Han Vinck
University of Essen, Germany

At the occasion of the 60th Birthday of Paddy Farrell

Abstract. We describe several attempts to reduce the decoding complexity for convolutional codes. Although most of the attempts were unsuccessful, we gained interesting insight information which helped in understanding the theory of convolutional codes and led to further studies in the field. We conclude that the performance of the standard constraint length 6 Viterbi decoder is hard to beat. We give a concatenated coding scheme using the same code and a single parity check code to obtain an almost ML decoding algorithm with performance of a free distance 20 code. We also give several interesting applications of the dual encoder. We show that the concept of the dual encoder is of practical as well as of theoretical importance.

1. Syndrome Decoding [1,2]

In syndrome decoding we use the fact that if there are less relevant noise sequences than information sequences, it should be easier to build a decoder based on the decoding of noise instead of the information sequence. However, the decoding state space based on the syndrome former has the same size as the one based on the encoder. Complexity reduction is possible by 1) Restriction on the number of metric combinations. The question arises whether codes can be constructed that give rise to short metric combination tables. It is known that the number of metric combinations for a syndrome former based decoder is far less than the number of metric combinations for the encoder based decoder. In 1974, we decribed a simple ROM based implementation for a constraint length 2 or 4-state decoder, [11]. This kind of implementation can only be used for short constraint length codes or with the combination of special codes. 2) Reduction of the number of states to be considered. We introduce a class of codes with reduced state space complexity. However, due to the requested symmetries the distance decreases, see [2].

The M-algorithm can also be used to do a reduced state space decoding. However, up to now, all results reported do not beat the classical Viterbi decoder in decoding performance.

2. Classes of codes [3,4]

In [3], we introduce a class of R = ½ codes and derive some of the properties. The class of codes is characterized by the same μ, $\mu > 1$ initial coefficients for both encoder connection polynomials. The class of codes is used to improve the performance for sequential decoding. It allows the extension of equivalence classes of states. The same class of codes can also be used to improve the performance of the M-algorithm, where only a few number of estimates is considered, [4]. The M-algorithm can be implemented using the medium search principle and clever path register scheduling.

3. Concatenated codes [5,6]

The idea is to use the fact that when two codes are concatenated, the overall distance is the product of the individual codes. The problem is how to use this increase at low complexity.

Encoding: The encoding consists of N convolutional encoders. The Nth code sequence is the modulo two sum of the other N-1 sequences.

Decoding: First independently decode each of the received N convolutional code sequences with a regular Viterbi ML decoder. Then, if error events occur, determine the particular Viterbi decoder that caused this error event. If no more than one Viterbi decoder is in error, the post decoding gives an ML estimate. Considering N = 12, constraint length $\nu=2$ encoders, we achieve a performance comparable to the constraint length 6, 64 state decoding complexity, code. However, the complexity is proportional to a 4 state decoder. Furthermore, we achieve a high degree (12) of parallelism. The same scheme can be used to develop a free distance 20 decoder with low complexity and high speed.

The same principle has been applied to Reed-Solomon-SPC coding with soft decision, [6]. We also considered application of the method to intersymbol interference channels, [7].

The main idea for all the above algorithms is that we do not correct isolated symbols, but look at error events instead. If in the concatenated code, only one decoder is in error, we have an ML decoding algorithm with low complexity. Degradation to multiple error events is shown to be negligible.

4. Sequential Decoding for High Rate Codes [8]

For high rate (N-1)/N codes, we give an improvement of the classical STACK decoding algorithm. The average number of stack reorganizations as well as the number of successors per extension can be reduced considerably, without increase of decoding error probability. The savings are based on the following:

1. Instead of storing 2^{N-1} successors per extended estimate, we are able to store only a small fraction of this number.
2. In the error free case, no estimate needs to be stored in the stack, and hence, no stack reorganization takes place.

In conclusion, we can state that this implementation makes the stack decoder more attractive to implement than the Fano decoder.

5. The dual encoder [9, 10]

The encoder plays an important role in the decoding for convolutional codes. The dual code plays the role of a „free gift“. Nice to have, but not very useful. We show that the dual code has many applications. We used the dual encoder in the development of an algorithm for identifying rate (n-1)/n catastrophic punctured convolutional encoders. The algorithm first finds a nonzero codeword of the dual of the puntured rate (n-1)/n code. Since the dual code is a rate 1/n code, its minimum encoder can be easily found from any nonzero codeword. The constraint length of a minimum encoder is always equal to that of the minimum encoder of its dual. Hence, the minimality of the punctured encoder, and thus the catastrophic property is determined. The algorithm is computationally efficient for both large and small constraint lengths.

Another surprising area where the dual code can be used, is the correction of defects. In [10], we show how convolutional codes may be implemented with a modest complexity. The probability of not being able to match a particular defect depends on the minimum weight w(t) of the dual codewords that leave the all-zero state and do not return to it within t steps.

We recall only one Theorem that shows an interesting relation between encoder and dual encoder. Let G_0 and H_0 be the respective dual minimal encoders. The obvious realization of G_0 involves k shift registers of length ν_i, $1 \le i \le k$. The adjoint-obvious realization of H_0 involves n-k shift registers of length $\nu_i^{\perp}$, $1 \le i \le n\text{-}k$. Forney shows that $\sum\nu_i = \sum\nu_i^{\perp}$ and that the respective state spaces are isomorphic. Define $\nu_{max} = \max_{1 \le i \le k} \nu_i$ and $\nu_{min}^{\perp} = \min_{1 \le i \le n-k} \nu_i^{\perp}$. The following theorem shows that the encoder G_0 can generate a sequence that matches bursts of defects of n-tuples if preceded by a defect free period, or guard space, of ν_{max} n-tuples.

Theorem. The encoder G_0 can generate any output segment of n-tuples of length $\nu_{min}^{\perp}$ with respect to a preceding guard space of ν_{max} n-tuples.

6. Conclusions

In this contribution to the birthday of Paddy Farrell, we consider some applications of convolutional codes and the related complexity. We show that the concept of the dual encoder is of practical as well as of theoretical importance. We also give some examples of low complexity decoding algorithms.

References

[1] J.P.M. Schalkwijk, A.J. Vinck and K.A. Post, „Syndrome Decoding of binary k/n Convolutional Codes," IEEE Trans. on Comm. pp. 977-985, Sept. 1976.

[2] A.J. Vinck, „Syndrome Decoding of Binary Convolutional Codes," Ph.D. thesis 1980, Technical university of Eindhoven, the Netherlands.

[3] A.J. Vinck, A.J.P. de Paepe, J.P.M. Schalkwijk, „A Class of Binary Rate One-Half Convolutional Codes that Allows an Improved Stack Decoder," IEEE Tr. Inform. Theory., Vol. IT-26, July 1980, pp. 389-392.

[4] A.J. Vinck and R. van der Vleuten, „A Method to Implement Linear Complexity Convolutional Decoder," Benelux Symposium on Information Theory, 1988, pp. 75-80.

[5] R. Schweikert and A.J. Vinck, „A Convolutional Single-Parity-Check Concatenated Coding Scheme for High-Data-Rate Applications," IEEE Tr. on Comm., Vol. 39, Jan 1991, pp. 4-7.

[6] F. Dolainsky, R. Schweikert, A.J. Vinck, „ A Concatenated Reed-Solomon-SPC coding Scheme with Soft Decision," Space communications 7, 1990, pp. 293-297.

[7] M. Schnell and A.J. Han Vinck, „A Low complexity Equalization Concept for Frequency-Selective Mobile Radio channels,“ AEÜ Int. J. Electron. Commun., Bd 50, 1996, pp. 343-352.
[8] A.J. Vinck, „A Low Complexity Stack Decoder for a Class of Binary Rate (n-1)/n convolutional Codes,“ IEEE Tr. on Comm., Vol. 32, pp. 476-479, 1984.
[9] Feng-Wen Sun and A.J. Han Vinck, „An Algorithm for Identifying Rate (n-1)/n catastrophic Punctured Convolutional Encoders,“ IEEE trans. on Information theory, Vol. 42, May 1996, pp. 1010-1013.
[10] J. Martin Borden and A.J. Vinck, „On coding for „Stuck-At“ Defects,“ IEEE Trans. on Information Theory, Vol. IT-33, September 1987, pp. 729-735.
[11] J.P.M. Schalkwijk and A.J. Vinck, „Syndrome Decoding of Convolutional Code,“ IEEE Trans. on Comm., Vol. Com-23, pp. 789-792, July 1975.

12. The Decoding of Binary Quasi-Cyclic Codes

Petra Heijnen and Henk van Tilborg*

October 31, 1997

Abstract

Many binary quasi-cyclic codes have a better minimum distance than other known linear codes with the same parameters. Therefore a lot of effort has been put into finding such codes. Until now, practically nothing was known about the decoding of quasi-cyclic codes. This paper is concerned with the decoding of binary quasi-cyclic codes. Furthermore a universal decoding algorithm for such codes will be presented.

1 Quasi-cyclic codes

For a q-ary vector $\mathbf{a}$ of length k, we denote by $\Lambda\mathbf{a}$ the cyclic shift of $\mathbf{a}$ over one position.
Consider a q-ary $[mk, k, d]$ code C. Let $\mathbf{c}$ be a codeword of C. We divide $\mathbf{c}$ in m parts, say $\mathbf{c} = (\mathbf{c}_0, \mathbf{c}_1, \dots, \mathbf{c}_{m-1})$, each of length k.

Definition 1 *C is called a quasi-cyclic (QC) code if and only if for every word* $\mathbf{c}$

$$\mathbf{c} = (\mathbf{c}_0, \mathbf{c}_1, \dots, \mathbf{c}_{m-1}) \in C \Rightarrow (\Lambda\mathbf{c}_0, \Lambda\mathbf{c}_1, \dots, \Lambda\mathbf{c}_{m-1}) \in C.$$

An $[mk, k]$ QC-code C has a generator-matrix G that consists of m circulant $k \times k$ matrices $G_0, G_1, \dots, G_{m-1}$,

$$G = \left(\begin{array}{c|c|c|c} G_0 & G_1 & \dots & G_{m-1} \end{array} \right).$$

A $k \times k$ circulant matrix P has the following form:

$$P = \begin{pmatrix} p_0 & p_1 & \cdots & p_{k-1} \\ p_{k-1} & p_0 & \cdots & p_{k-2} \\ \vdots & \ddots & \ddots & \vdots \\ p_1 & \cdots & p_{k-1} & p_0 \end{pmatrix}.$$

We call the polynomial

$$p(x) = \sum_{i=0}^{k-1} p_i x^i.$$

the associated polynomial of the circulant matrix P.
Similarly, for the matrix G_i, we call its associated polynomial $g_i(x)$, $0 \le i \le m-1$.
We represent the m parts of a codeword $\mathbf{c}$ by m polynomials, say $c_i(x)$, $0 \le i \le m-1$, of degree less than k. So, we will also denote $\mathbf{c}$ by $\mathbf{c}(x) = (c_0(x), c_1(x) \dots, c_{m-1}(x))$. We will use both notations.
Now, $\mathbf{c}(x)$ is a codeword in C if there exists a polynomial $a(x)$ of degree less than k, such that $c_i(x) \equiv a(x)g_i(x) \mod x^k - 1$, for all i, $0 \le i \le m-1$. Therefore we call the polynomials

*Department of Mathematics and Computing Science, Eindhoven University of Technology, P.O. Box 513, 5600 MB Eindhoven, The Netherlands

$g_0(x), g_1(x), \ldots, g_{m-1}(x)$ the *generator-polynomials* of C.

To be able to derive properties of QC codes we will need the following well-known lemma on linear congruence relations for polynomials.

Lemma 1 *Let $g(x)$, $c(x)$ and $f(x)$ be polynomials with coefficients in $\mathbb{F}_q$. Then*

$$a(x)g(x) \equiv c(x) \mod f(x) \tag{1}$$

has a solution for $a(x)$ if and only if $\gcd(g(x), f(x))|c(x)$.
If so, then there are exactly $T = q^{\deg(\gcd(g(x),f(x)))}$ solutions. If $a_0(x)$ is one such solution then the others can be written as

$$a_0(x) + t(x)\frac{f(x)}{\gcd(g(x), f(x))} \mod f(x),$$

for $t(x)$ with $\deg(t(x)) < \deg(\gcd(g(x), f(x)))$.

Theorem 1 *Let G be the generator-matrix of a q-ary $[mk, k]$ QC code, generated by the polynomials $g_0(x), g_1(x), \ldots, g_{m-1}(x)$, then*

$$\text{rank}(G) = k - \deg(\gcd(g_0(x), g_1(x), \ldots, g_{m-1}(x), x^k - 1)).$$

Proof Let $a(x)$ be a nonzero polynomial of degree $< k$. Then by Lemma 1

$$a(x)g_i(x) \equiv 0 \mod x^k - 1, \text{ for all } 0 \leq i \leq m-1, \tag{2}$$

if and only if

$$\frac{x^k - 1}{\gcd(g_i(x), x^k - 1)} \text{ divides } a(x), \text{ for all } 0 \leq i \leq m-1,$$

i.e. if and only if

$$f(x) = \frac{x^k - 1}{\gcd(g_0(x), \ldots, g_{m-1}(x), x^k - 1)} \text{ divides } a(x).$$

It follows that $\text{rank}(G) = k - \dim\{\mathbf{a}|\mathbf{a}G = \mathbf{0}\} = k - (k - \deg(f(x))) =$
$k - \deg(\gcd(g_0(x), \ldots, g_{m-1}(x), x^k - 1))$.

■

Corollary 1 *(See [4])*

$$\text{rank}(G) = k \textit{ if and only if } \gcd(g_0(x), \ldots, g_{m-1}(x), x^k - 1) = 1.$$

Theorem 2 *Let C be the q-ary $[mk, k]$ QC code, generated by $g_0(x), \ldots, g_{m-1}(x)$. Then $\mathbf{c}(x) \in C$ if and only if*

$$c_i(x)g_j(x) + c_j(x)g_i(x) \equiv 0 \mod x^k - 1, \textit{ for all } i, j \textit{ with } 0 \leq i, j \leq m-1. \tag{3}$$

Proof $c(x) \in C$ if and only if there exists a polynomial $a(x)$, $\deg(a(x)) < k$, such that $c_i(x) \equiv a(x)g_i(x) \mod x^k - 1$, $0 \leq i \leq m-1$.
($\Rightarrow$) Filling this in into (3) will prove the theorem from left to right.
($\Leftarrow$) From Corollary 1 we know that $\gcd(g_0(x), \ldots, g_{m-1}(x), x^k - 1) = 1$.
Define $\gamma_i(x) = \gcd(g_i(x), x^k - 1)$, $0 \leq i \leq m-1$. Then

$$\gcd(\gamma_0(x), \ldots, \gamma_{m-1}(x)) = 1.$$

By the Euclidean Algorithm we can find polynomials $\delta_0(x), \ldots, \delta_{m-1}(x)$, such that

$$\delta_0(x)\gamma_0(x) + \ldots + \delta_{m-1}(x)\gamma_{m-1}(x) = 1.$$

Similarly, we can find polynomials $h_i(x)$ such that

$$h_i(x)g_i(x) \equiv \gamma_i(x) \mod x^k - 1,\ 0 \le i \le m-1.$$

Take

$$a(x) \equiv c_0(x)h_0(x)\delta_0(x) + \ldots + c_{m-1}(x)h_{m-1}(x)\delta_{m-1}(x) \mod x^k - 1.$$

We shall show that $a(x)$ satisfies $c_i(x) \equiv a(x)g_i(x) \mod x^k - 1$ for all i, $0 \le i \le m-1$.
From (3) we have $c_j(x)g_i(x) \equiv c_i(x)g_j(x) \mod x^k - 1$, $0 \le j \le m-1$. So we get

$$\begin{aligned} a(x)g_i(x) &\equiv c_0(x)h_0(x)\delta_0(x)g_i(x) + \ldots + c_i(x)h_i(x)\delta_i(x)g_i(x) + \ldots + c_{m-1}(x)h_{m-1}(x)\delta_{m-1}(x)g_i(x) \\ &\equiv c_i(x)h_0(x)g_0(x)\delta_0(x) + \ldots + c_i(x)h_i(x)g_i(x)\delta_i(x) + \ldots + c_i(x)h_{m-1}(x)g_{m-1}(x)\delta_{m-1}(x) \\ &\equiv c_i(x)(\gamma_0(x)\delta_0(x) + \ldots + \gamma_i(x)\delta_i(x) + \ldots + \gamma_{m-1}(x)\delta_{m-1}(x)) \\ &\equiv c_i(x) \mod x^k - 1,\ 0 \le i \le m-1. \end{aligned}$$

So there exists an $a(x)$ such that $c_i(x) \equiv a(x)g_i(x) \mod x^k - 1$ and thus

$$\mathbf{c}(x) = (c_0(x), \ldots, c_{m-1}(x)) \in C.$$

■

We call the equations in (3) the *parity check equations* of the q-ary QC code C. Note that not all those equations are independent. For a fixed i they are independent but in general not sufficient.

Computer evidence suggests that QC codes with good parameters always have at least one of the circulant matrices G_i, $0 \le i \le m-1$, of full rank. In that case we can find an equivalent QC code of C, where G_0 is the $k \times k$ identity-matrix, i.e. its associated polynomial is 1. From now on we assume that $g_0(x) = 1$.

2 Decoding

We will now describe a bounded distance decoding algorithm for binary QC codes. It makes use of the parity check equations of Theorem 2.
Let C be a binary $[mk, k, d]$ QC code generated by $g_0(x) = 1, g_1(x), \ldots, g_{m-1}(x)$. We shall first consider the special case that

$$\gcd(g_i(x), x^k - 1) = 1,\ \forall i,\ 0 \le i \le m-1.$$

2.1 Syndrome decoding when all circulants have full rank

In the case that all $g_i(x)$ are coprime with $x^k - 1$, one can find by the Euclidean algorithm, polynomials $g_i^{-1}(x)$ such that

$$g_i^{-1}(x)g_i(x) \equiv 1 \mod x^k - 1,\ 0 \le i \le m-1.$$

For a certain range of parameters, QC codes that satisfy this condition are listed in Table 2 in Appendix A.
In the sequel we assume that $\mathbf{c}(x) = (c_0(x), \ldots, c_{m-1}(x)) \in C$ is transmitted, while $\mathbf{r}(x) = (r_0(x), \ldots, r_{m-1}(x))$ is received. Let $\mathbf{e}(x) = (e_0(x), \ldots, e_{m-1}(x))$ represent the error-vector, so $r_i(x) = c_i(x) + e_i(x)$, for $0 \le i \le m-1$.
The number of errors that C can correct is $t = \lfloor \frac{d-1}{2} \rfloor$. We assume that the weight of the error-vector $\mathbf{e}$ is not more than t.
Theorem 2 gives the parity check equations for C.

$$\mathbf{c}(x) \in C \Leftrightarrow c_i(x)g_j(x) + c_j(x)g_i(x) \equiv 0 \mod x^k - 1,\ 0 \le i, j \le m-1.$$

Since $\gcd(g_i(x), x^k - 1) = 1$, $0 \le i \le m-1$, we can rewrite this as

$$c_j(x) + c_i(x)g_j(x)g_i^{-1}(x) \equiv 0 \mod x^k - 1,\ 0 \le i, j \le m-1.$$

Take i fixed. Define the syndrome-polynomials $s_j^{(i)}(x)$ by

$$s_j^{(i)}(x) \equiv r_j(x) + r_i(x)g_j(x)g_i^{-1}(x) \mod x^k - 1,\ 0 \le j \le m-1.$$

It is easy to check that

$$s_j^{(i)}(x) \equiv e_j(x) + e_i(x)g_j(x)g_i^{-1}(x) \mod x^k - 1,\ 0 \le j \le m-1$$

and that $s_i^{(i)}(x) \equiv 0 \mod x^k - 1$.

Now take an arbitrary polynomial $\varepsilon(x)$ of degree less than k and define the polynomial $\sigma_j(x)$ by

$$\sigma_j(x) \equiv \varepsilon(x)g_j(x)g_i^{-1}(x) \mod x^k - 1,\ 0 \le j \le m-1.$$

If $e_i(x) = \varepsilon(x)$ then $e_j(x) = s_j^{(i)}(x) + \sigma_j(x),\ 0 \le j \ne i \le m-1$.
If $e_i(x) \equiv x^l\varepsilon(x) \mod x^k - 1$, for a certain l, $0 \le l \le k-1$, i.e. $\boldsymbol{\varepsilon}$ is a cyclic shift of $\mathbf{e}_i$, then $e_j(x) \equiv s_j^{(i)}(x) + x^l\sigma_j(x) \mod x^k - 1,\ 0 \le j \ne i \le m-1$. We will make use of this fact when we try to decode $\mathbf{r}(x)$.

First of all we need some new definitions.
Consider all binary vectors of length k and partition them in disjunct classes, such that each class consists of a vector with all its cyclic shifts.
We call $\mathbf{y}$ the *cyclic representative* of a class Γ if $\mathbf{y}$ is the lexicographic smallest vector in Γ.

A vector s is called a *QC-shift* of a vector $\boldsymbol{\sigma}$, both of length pk, if for some l, $\mathrm{s} = (\Lambda^l\boldsymbol{\sigma}_0, \ldots, \Lambda^l\boldsymbol{\sigma}_p)$. We denote this by $\mathrm{s} = \Lambda^l_{\mathrm{qc}}\boldsymbol{\sigma}$.

A vector s is called a *distance t QC-shift* of a vector $\boldsymbol{\sigma}$, both of length pk, for some l, $0 \le l \le k-1$, $\mathrm{d}(\mathrm{s}, \Lambda^l_{\mathrm{qc}}\boldsymbol{\sigma}) \le t$.

Now we can decode $\mathbf{r}(x)$ in the following way.
Consider the binary vectors $\mathrm{s}^{(i)} = (\mathrm{s}_0^{(i)}, \ldots, \mathrm{s}_{i-1}^{(i)}, \mathrm{s}_{i+1}^{(i)}, \ldots, \mathrm{s}_{m-1}^{(i)})$, $0 \le i \le m-1$. If $e_i(x) = 0$ for some i, i.e. block i is errorfree, then $s_j^{(i)}(x) \equiv e_j(x) \mod x^k - 1,\ 0 \le j \ne i \le m-1$.
Therefore, if $\mathrm{wt}(\mathrm{s}^{(i)}) \le t$ for some i, we may conclude that block i was indeed errorfree, because we assume that no more than t errors have been made and since in that case $\mathrm{s}^{(i)} = (\mathbf{e}_0, \ldots, \mathbf{e}_{i-1}, \mathbf{e}_{i+1}, \ldots, \mathbf{e}_{m-1})$, we get $\mathrm{wt}(\mathrm{s}^{(i)}) = \mathrm{wt}(\mathbf{e})$. The most likely transmitted message-polynomial $a(x)$ is then given by

$$a(x) = (a(x)g_0(x) = c_0(x) = r_0(x) + e_0(x) =) \, r_0(x) + s_0^{(i)}(x).$$

So, we have now checked if one of the blocks was error-free. If not then we will look for a block that contains a small number of errors.
Let $\tau = \lfloor \frac{t}{m} \rfloor$. Then at least one of the error-vectors $\mathbf{e}_i$, $0 \le i \le m-1$, has weight $\le \tau$.

If the message $a(x)$ is not found yet, then we generate the cyclic representatives of weight $1, 2, \ldots, \tau$, by an algorithm that will be explained in Section 2.2. At least one of the error-vectors $\mathbf{e}_i$ will be a cyclic shift of one of those cyclic representatives.

Let $\boldsymbol{\varepsilon}$ be such a cyclic representative of weight w, $1 \le w \le \tau$. We will now check if one of the error-vectors $\mathbf{e}_i$ is a cyclic shift of $\boldsymbol{\varepsilon}$. To reduce the complexity of the computations we will first check if $\mathbf{e}_0$ is a cyclic shift of $\boldsymbol{\varepsilon}$. If not we can use the computation of $\sigma_j(x) \equiv \varepsilon(x)g_j(x) \mod x^k - 1$ for the other blocks.
Compute the polynomials

$$\sigma_j(x) \equiv \varepsilon(x)g_j(x) \mod x^k - 1,\ 0 \le j \le m-1.$$

We check if $\boldsymbol{\sigma} = (\boldsymbol{\sigma}_1, \ldots, \boldsymbol{\sigma}_{m-1})$ is a distance $(t-w)$ QC-shift of $\mathbf{s}^{(0)} = (\mathbf{s}_1^{(0)}, \ldots, \mathbf{s}_{m-1}^{(0)})$.
If so, say $\mathrm{d}(\mathbf{s}^{(0)}, \Lambda_{\mathrm{qc}}^{l}\boldsymbol{\sigma}) \leq t-w$, then the most likely transmitted message-polynomial is,

$$a(x) \equiv \Big(r_0(x) + e_0(x) \equiv r_0(x) + s_0^{(0)}(x) + x^l\sigma_0(x) \equiv x^l\varepsilon(x)g_0(x) + r_0(x) = \Big)\, x^l\dot{\varepsilon}(x) + r_0(x) \mod x^k - 1.$$

If not, check for $i = 1, 2, \ldots, m-1$ if $\mathbf{e}_i$ is a cyclic shift of $\boldsymbol{\varepsilon}$. Therefore, compute for $i = 1, 2, \ldots, m-1$ the polynomial-vectors

$$\boldsymbol{\sigma}^{(i)}(x) = (\sigma_0(x)g_i^{-1}(x), \ldots, \sigma_{i-1}(x)g_i^{-1}(x), \sigma_{i+1}(x)g_i^{-1}(x), \ldots, \sigma_{m-1}(x)g_i^{-1}(x)),$$

and check if $\boldsymbol{\sigma}^{(i)}$ is a distance $(t-w)$ QC-shift of $\mathbf{s}^{(i)} = (\mathbf{s}_0^{(i)}, \ldots, \mathbf{s}_{i-1}^{(i)}, \mathbf{s}_{i+1}^{(i)}, \ldots, \mathbf{s}_{m-1}^{(i)})$.
If so, say $\mathrm{d}(\mathbf{s}^{(i)}, \Lambda_{\mathrm{qc}}^{l}\boldsymbol{\sigma}^{(i)}) \leq t-w$, then the most likely transmitted message-polynomial is,

$$a(x) \equiv \Big(r_0(x) + e_0(x) \equiv r_0(x) + s_0^{(i)}(x) + x^l\sigma_0(x) \equiv\Big)\, x^l\sigma_0^{(i)}(x) + s_0^{(i)}(x) + r_0(x) \mod x^k - 1.$$

To reduce the complexity of the computations we compute the syndrome-polynomials $s_j^{(i)}(x)$, $0 \leq i, j \leq m-1$, by $s_j^{(i)}(x) \equiv r_j(x) + r_i(x)g_j(x)g_i^{-1}(x) \mod x^k - 1$ for $0 \leq j < i \leq m-1$ and $s_j^{(i)}(x) \equiv s_i^{(j)}(x)g_j(x)g_i^{-1}(x) \mod x^k - 1$ for $0 \leq i < j \leq m-1$.

Let $T(k, w)$ be the total number of cyclic representatives of length k and weight w.
The decoding algorithm consists now of the following steps.

Algorithm 1

1. *Compute* $s_j^{(i)}(x) \equiv r_j(x) + r_i(x)g_j(x)g_i^{-1}(x) \mod x^k - 1,\ 0 \leq j < i \leq m-1.$
2. *Compute* $s_j^{(i)}(x) \equiv s_i^{(j)}(x)g_j(x)g_i^{-1}(x) \mod x^k - 1,\ 0 \leq i < j \leq m-1.$
3. *For i from 0 to* $m-1$ *do*
   If $\mathrm{wt}(\mathbf{s}^{(i)}) \leq t$ *then*

   $$a(x) = r_0(x) + s_0^{(i)}(x).$$
   Stop.

4. *For* w *from 1 to* $\tau = \lfloor \frac{t}{m} \rfloor$ *do*
   For α *from 1 to* $T(k, w)$ *do*
   (a) *Generate the next cyclic representative of weight* w*, say* $\boldsymbol{\varepsilon}$*. (See Algorithm 2)*
   (b) *Compute* $\sigma_j(x) \equiv \varepsilon(x)g_j(x) \mod x^k - 1,\ 0 \leq j \leq m-1.$
   (c) *If* $\mathrm{d}(\mathbf{s}^{(0)}, \Lambda_{qc}^{l}\boldsymbol{\sigma}) \leq t-w$ *for some value of* l*, then*

   $$a(x) \equiv x^l\varepsilon(x) + r_0(x) \mod x^k - 1.$$
   Stop.

   (d) *For i from 1 to* $m-1$ *do*
      i. *Compute* $\sigma_j^{(i)}(x) \equiv \sigma_j(x)g_i^{-1}(x) \mod x^k - 1,\ 0 \leq j \neq i \leq m-1.$
      ii. *If* $\mathrm{d}(\mathbf{s}^{(i)}, \Lambda_{qc}^{l}\boldsymbol{\sigma}^{(i)}) \leq t-w$ *for some value of* l*, then*

      $$a(x) \equiv x^l\sigma_0^{(i)}(x) + s_0^{(i)}(x) + r_0(x) \mod x^k - 1.$$
      Stop.

Note that a lot of these steps can be done in parallel.

2.2 Generating cyclic representatives of fixed weight

In this section, an algorithm is given that generates the cyclic representatives of length k with a certain weight w, $0 \leq w \leq k$. The algorithm is based on an algorithm from [1], that generates all cyclic representatives in lexicographic order.

We will need the following lemma.

Lemma 2 *K is a cyclic representative $\Leftrightarrow$*
If P is a prefix of K then K contains no subsequence of the same length as P that is lexicographic smaller than P.

Proof ($\Rightarrow$) If K contains a subsequence Q of the same length as a prefix P of K, such that Q is lexicographic smaller than P then the cyclic shift of K that starts with Q is lexicographic smaller than K itself.
($\Leftarrow$) If K contains no subsequence of the same length as a prefix P of K that is lexicographic smaller than K then K has no cyclic shift that is lexicographic smaller than K. Therefore K is the smallest cyclic shift of itself and K is a cyclic representative.

■

First of all we determine the number of cyclic representatives of weight w.
Define the *period* $p(\mathbf{v})$ of a vector $\mathbf{v}$ as the smallest positive value l such that $\mathbf{v} = \Lambda^l \mathbf{v}$. Let $\mathbf{x}$ be the prefix of $\mathbf{v}$ of length l. Then $\mathbf{v}$ consists of $k/p(\mathbf{v})$ repetitions of $\mathbf{x}$

Let $\mathcal{N}(k,w)$ be the set of all vectors of length k and weight w. Define $T(k,w)$ as the number of cyclic representatives in $\mathcal{N}(k,w)$ and $M(k,w)$ as the number of cyclic representatives in $\mathcal{N}(k,w)$ of period k.

Theorem 3
$$T(k,w) = \frac{1}{k} \sum_{d|\,\gcd(k,w)} \phi(d) \binom{k/d}{w/d},$$
where ϕ is Eulers Totient function (see [3]).

Proof Let $\mathbf{v} \in \mathcal{N}(k,w)$ with period d. Let $\mathbf{x}$ be the prefix of $\mathbf{v}$ of length d. Then $\mathbf{x}$ has period d. Also $\mathbf{v}$ consists of $b := k/d$ repetitions of $\mathbf{x}$. Let a be the weight of $\mathbf{x}$. So it follows that $b|\gcd(k,w)$ and that

$$T(k,w) = \sum_{b|\,\gcd(k,w)} M(k/b, w/b). \tag{4}$$

Since $\mathbf{v}$ has d distinct cyclic shifts,

$$\binom{k}{w} = \sum_{b|\,\gcd(k,w)} \frac{k}{b} M(k/b, w/b).$$

We will apply Möbius inversion (see [3]) on this equation to determine $M(k,w)$ explicitly. Take $\gamma = \gcd(k,w)$, $k = \gamma x$ and $w = \gamma y$. Define the functions $g(\gamma)$ by $\binom{\gamma x}{\gamma y}$ and $f(b)$ by $f(b) = bxM(bx, by)$. Then

$$\begin{aligned} g(\gamma) = \binom{\gamma x}{\gamma y} &= \sum_{b|\gamma} \frac{\gamma x}{b} M(\gamma x/b, \gamma y/b) \\ &\overset{\gamma/b \to b}{=} \sum_{b|\gamma} bxM(bx, by) = \sum_{b|\gamma} f(b) \end{aligned}$$

Applying Möbius inversion we obtain

$$f(\gamma) = \sum_{b|\gamma} \mu(b) g(\gamma/b),$$

where μ is the Möbius function. This yields

$$\gamma x M(\gamma x, \gamma y) = \sum_{b|\gamma} \mu(b) \binom{\gamma x/b}{\gamma y/b}.$$

If we translate this back to k and w, we get

$$M(k,w) = \frac{1}{k} \sum_{b|gcd(k,w)} \mu(b) \binom{k/b}{w/b}.$$

We substitute this into (4) and get

$$T(k,w) = \frac{1}{k} \sum_{a|\gcd(k,w)} a \sum_{b|\gcd(k/a,w/a)} \mu(b) \binom{k/ab}{w/ab},$$

which can be simplified by changing the order of summation and putting $d = ab$, into

$$T(k,w) = \frac{1}{k} \sum_{d|\gcd(k,w)} \binom{k/d}{w/d} \sum_{a|d} a\mu(d/a).$$

From [3] we get $\sum_{a|d} a\mu(\frac{d}{a}) = \sum_{a|d} \frac{d}{a}\mu(a) = \phi(d)$, the Euler function of d. So

$$T(k,w) = \frac{1}{k} \sum_{d|\gcd(k,w)} \phi(d) \binom{k/d}{w/d}.$$

■

Example 1 *Compute the number of cyclic representatives of length 12 and weight 6.*

$$T(12,6) = \frac{1}{12}\left(\phi(1)\binom{12}{6} + \phi(2)\binom{6}{3} + \phi(3)\binom{4}{2} + \phi(6)\binom{2}{1}\right) = 80$$

We shall now describe an algorithm that determines the cyclic representatives of length k and weight w. This algorithm will generate a list of candidates in lexicographic order. Most of those candidates will be cyclic representatives, but not all of them. Every new candidate is generated from the previous one. The algorithm starts with the lexicographic smallest vector of length k and weight w, i.e. $[00\ldots0\underbrace{11\ldots1}_{w}]$.

Assume that K_L was the last candidate of weight w, that the algorithm has found. This K_L has in general the form $[\mathbf{x}, 0, 1, \ldots, 1, 0, \ldots, 0]$. (The number of zeroes at the right can be zero) When searching for a lexicographic larger candidate of weight w it has no use to replace the zeroes at the right by ones, since the weight of the new vector becomes larger than w in that way. So the next candidate K_N will have $P = [\mathbf{x}, 1]$ as prefix. By Lemma 2, if K_N is a cyclic representative then K_N cannot contain a subsequence of the same length as P that is lexicographic smaller than P. On the other hand, if we choose a K_N, that contains a subsequence of the same length as the prefix P that is lexicographic larger than P, then it is possible that we miss some cyclic representatives of weight w. The new candidate K has therefore the following form.

$$K = [P, P, \ldots, P, \overline{P}],$$

where $\overline{P}$ is a prefix of P such that K has length k. We distinguish four cases.

1. If the weight of K is smaller than w, then we replace, starting at the right, zeroes by ones, until K has weight w. This new vector is K_N, it contains no subsequences of the same length as P, that are lexicographic smaller than P. Therefore by Lemma 2 it is a cyclic representative of weight w.
2. If the weight of K equals w and $\overline{P}$ is not equal to P, then K has a cyclic shift that is lexicographic smaller than K. K is the new candidate K_N, but by Lemma 2 it is not a cyclic representative of weight w.
3. If the weight of K equals w and $\overline{P}$ equals P then K is the new candidate K_N and by Lemma 2 it is a cyclic representative of weight w.
4. If the weight of K is larger than w, then K is the new candidate K_N, but it is not a cyclic representative of weight w.

Now we can write down the final form of the algorithm that generates in lexicographic order the cyclic representatives of length k and weight w.

Algorithm 2

1. $P := [0]$, $i := 0$.

2. While $i < T(k,w)$ *do*

(a) $K := [P, P, \dots, P, \overline{P}]$.

(b) If $\mathrm{wt}(K) = w$ *and* $\overline{P} = P$, *then* K *is a cyclic representative.* $i := i + 1$.

(c) If $\mathrm{wt}(K) < w$ *then replace* $(w - \mathrm{wt}(K))$ *zeroes by ones at the right of* K. *This new* K *is a cyclic representative.* $i := i + 1$.

(d) Determine $\mathbf{x}$ *such that* $K = [\mathbf{x}, 0, 1, 1, \dots, 1, 0, 0 \dots, 0]$ *and set* $P := [\mathbf{x}, 1]$.

Example 2 *Determine all cyclic representatives of length 9 and weight 5. There are* $T(9,5) = \frac{1}{9}\binom{9}{5} =$ *14 cyclic representatives, which are printed boldface.*

i	P	$[P, \dots, P, \overline{P}]$	K
0	0	000000000	**0 0 0 0 1 1 1 1 1**
1	0001	000100010	**0 0 0 1 0 1 1 1 1**
2	00011	000110001	**0 0 0 1 1 0 1 1 1**
3	000111	000111000	**0 0 0 1 1 1 0 1 1**
4	0001111	000111100	**0 0 0 1 1 1 1 0 1**
5	00011111	000111110	0 0 0 1 1 1 1 1 0
	001	001001001	**0 0 1 0 0 1 1 1 1**
6	00101	001010010	**0 0 1 0 1 0 1 1 1**
7	001011	001011001	**0 0 1 0 1 1 0 1 1**
8	0010111	001011100	**0 0 1 0 1 1 1 0 1**
9	00101111	001011110	0 0 1 0 1 1 1 1 0
	0011	001100110	**0 0 1 1 0 0 1 1 1**
10	001101	001101001	**0 0 1 1 0 1 0 1 1**
11	0011011	001101100	**0 0 1 1 0 1 1 0 1**
12	00110111	001101110	0 0 1 1 0 1 1 1 0
	00111	001110011	0 0 1 1 1 0 0 1 1
	0011101	001110100	**0 0 1 1 1 0 1 0 1**
13	00111011	001110110	0 0 1 1 1 0 1 1 0
	001111	001111001	0 0 1 1 1 1 0 0 1
	00111101	001111010	0 0 1 1 1 1 0 1 0
	0011111	001111100	0 0 1 1 1 1 1 0 0
	01	010101010	**0 1 0 1 0 1 0 1 1**
14			

2.3 Not all the circulants have full rank

Assume that we want to use an $[mk, k, d]$ QC code and that we know that there exists a QC code with these parameters, but that we cannot find one with all the circulants of full rank.
Let C be a $[mk, k, d]$ QC code, generated by $g_0(x) = 1, g_1(x), \ldots, g_{m-1}(x)$, with ρ circulants of full rank. We can use Algorithm 1 with some minor changes. The consequences for the decoding complexity may be more serious when $e \ll m$.
We restrict ourselves to the circulants of full rank. Define the set V by $V := \{i \mid \gcd(g_i(x), x^k - 1) = 1,\ 0 \le i \le m-1\}$. In the worst case all the errors are in the blocks $i \in V$. In that case one of the error-vectors $\mathbf{e}_i$, $i \in V$ is of weight $\le \tau = \lfloor t/\rho \rfloor$. So we need to generate the cyclic representatives of length k and weight w, $1 \le w \le \tau$, in the worst case.

Algorithm 3

1. *Compute* $s_j^{(i)}(x) \equiv r_j(x) + r_i(x) g_j(x) g_i^{-1}(x) \mod x^k - 1,\ i \in V,\ 0 \le j \le m-1.$
2. *For* $i \in V$ *do*
   If $\mathrm{wt}(\mathbf{s}^{(i)}) \le t$ *then*

   $$a(x) = r_0(x) + s_0^{(i)}(x).$$

   Stop.

3. *For* w *from 1 to* $\tau = \lfloor \frac{t}{\rho} \rfloor$ *do*
   For α *from 1 to* $T(k, w)$ *do*
   (a) *Generate the next cyclic representative of weight* w, *say* $\boldsymbol{\varepsilon}$. *(See Algorithm 2)*
   (b) *Compute* $\sigma_j(x) \equiv \varepsilon(x) g_j(x) \mod x^k - 1,\ 0 \le j \le m-1.$
   (c) *If* $\mathrm{d}(\mathbf{s}^{(0)}, \Lambda_{qc}^l \boldsymbol{\sigma}) \le t - w$ *for some value of* l, *then*

   $$a(x) \equiv x^l \varepsilon(x) + r_0(x) \mod x^k - 1.$$

   Stop.

   (d) *For* $i \in V \backslash \{0\}$ *do*
      i. *Compute* $\sigma_j^{(i)}(x) \equiv \sigma_j(x) g_i^{-1}(x) \mod x^k - 1,\ 0 \le j \ne i \le m-1.$
      ii. *If* $\mathrm{d}(\mathbf{s}^{(i)}, \Lambda_{qc}^l \boldsymbol{\sigma}^{(i)}) \le t - w$ *for some value of* l, *then*

      $$a(x) \equiv x^l \sigma_0^{(i)}(x) + s_0^{(i)}(x) + r_0(x) \mod x^k - 1.$$

      Stop.

In Table 2 in Appendix A one can find for a certain range of parameters QC codes with a large number of circulants of full rank.

3 Complexity of the algorithm

We will count the number of operations in Algorithm 1 in the worst case. Therefore we distinguish between the following computations.

$\mathrm{PM}(k)$	:	Multiplication of two binary polynomials of degree $< k$.
$\mathrm{PA}(k)$	:	Addition of two binary polynomials of degree $< k$.
$\mathrm{WT}(n, t)$	:	Test if a vector of length n has weight $\le t$.
$\mathrm{CR}(k, w)$	:	Generation of all cyclic representatives of length k and weight w.
$\mathrm{DT}(n, t)$	:	Test if two vectors of length n are at distance $\le t$.
$\mathrm{QC}(m, k)$	:	Quasi-cyclic shift of a vector of length mk over one position.

Now we give the number of binary operations per computation.

Computation		# Binary operations
$\mathrm{PM}(k)$	:	k
$\mathrm{PA}(k)$	:	1
$\mathrm{WT}(n,t)$	:	n
$\mathrm{DT}(n,t)$	:	n
$\mathrm{QC}(m,k)$	:	m

In our examples k is relatively small, therefore addition of two polynomials can be viewed as a single operation. We assume that multiplication of two polynomials needs in the worst case k additions.
In the worst case one has to check all the bits of a vector to know if its weight exceeds t.
We assume that shifting a vector of length k cyclicly over one position, when k is not too large, is a single operation. A QC-shift of a vector of length mk, over one position needs therefore m binary operations.

The number of binary operations $\mathrm{CR}(k,w)$ that Algorithm 2 needs to generate all cyclic representatives of length k and weight w is listed in Table 3.
We count the computation of the weight of K in Step (2b) as k binary operations and testing if $\mathrm{wt}(K) = w$ as 1 binary operation. The same for testing if $\overline{P} = P$.
In Step (2c) say that we need to check N bits before we see $(\mathrm{wt}(k) - w)$ zeroes. We count this as N binary operations. For Step (2d) we need $(k - |\mathbf{x}|)$ binary operations. We can now compute the number of binary operations to generate all cyclic representatives of length k and weight w.

k	w	$\mathrm{CR}(k,w)$
5	1	7
6	1	8
6	2	42
6	3	56
7	1	9
8	1	10
8	2	92
9	1	11
9	2	96
10	1	12
10	2	165
11	1	13
11	2	180
11	3	600
12	1	14

k	w	$\mathrm{CR}(k,w)$
12	2	282
12	3	1045
13	1	15
13	2	294
13	3	1122
14	1	16
14	2	434
14	3	1768
15	1	17
15	2	455
15	3	2952
15	4	7644
16	1	18
16	2	640
16	3	2870

Table 1: Number of binary operations $\mathrm{CR}(k,w)$ that Algorithm 2 needs to generate all cyclic representatives of length k and weight w

The number of computations of the successive steps in Algorithm 1 is given by

Step 1 : $\frac{1}{2}m(m-1)(2\mathrm{PM}(k)+\mathrm{PA}(k))$

Step 2 : $\frac{1}{2}m(m-1)2\mathrm{PM}(k)$

Step 3 : $m\mathrm{WT}((m-1)k,t)$

Step 4ai : $\sum_{w=1}^{\lfloor t/m \rfloor}\mathrm{CR}(k,w)$

Step 4aii : $m\sum_{w=1}^{\lfloor t/m \rfloor}T(k,w)\mathrm{PM}(k)$

Step 4aiii : $k\sum_{w=1}^{\lfloor t/m \rfloor}T(k,w)(\mathrm{DT}((m-1)k,t-w)+\mathrm{QC}(m-1,k))$

Step 4aivA : $(m-1)^2\sum_{w=1}^{\lfloor t/m \rfloor}T(k,w)\mathrm{PM}(k)$

Step 4aivB : $(m-1)k\sum_{w=1}^{\lfloor t/m \rfloor}T(k,w)(\mathrm{DT}((m-1)k,t-w)+QC(m-1,k))$

For the total number of computations this yields:

$$\begin{array}{rl}
\left(2m(m-1)+(m^2-m+1)\sum_{w=1}^{\lfloor t/m \rfloor}T(k,w)\right) & \mathrm{PM}(k)+ \\
\frac{1}{2}m(m-1) & \mathrm{PA}(k)+ \\
m & \mathrm{WT}((m-1)k,t)+ \\
\sum_{w=1}^{\lfloor t/m \rfloor} & \mathrm{CR}(k,w)+ \\
mk\sum_{w=1}^{\lfloor t/m \rfloor}T(k,w) & \mathrm{DT}((m-1)k,t-w)+ \\
mk\left(\sum_{w=1}^{\lfloor t/m \rfloor}T(k,w)\right) & \mathrm{QC}(m-1,k).
\end{array}$$

We approximate $T(k,w)$ (See Theorem 3) by $\frac{1}{k}\binom{k}{w}$.

Now we can write the complexity of Algorithm 1 as the number $B_{\mathrm{all}}(k,m,\lfloor t/m \rfloor)$ of binary operations.
$B_{\mathrm{all}}(k,m,\lfloor t/m \rfloor)=$

$$m(m-1)(2\frac{1}{2}+k)+(m(m-1)(k+2)+1)\sum_{w=1}^{\lfloor t/m \rfloor}\binom{k}{w}+\sum_{w=1}^{\lfloor t/m \rfloor}\mathrm{CR}(k,w). \tag{5}$$

In the same way we can determine the complexity of Algorithm 3, when just ρ circulants are of full rank, and we get for the number of binary operations
$B_{\mathrm{some}}(m,k,\rho,\lfloor t/\rho \rfloor)=$

$$\rho(m-1)(3k+1)+((\rho+1)(m-1)(k+2)+1)\sum_{w=1}^{\lfloor t/\rho \rfloor}\binom{k}{w}+\sum_{w=1}^{\lfloor t/\rho \rfloor}\mathrm{CR}(k,w). \tag{6}$$

A trivial way to decode a received word is by just comparing all the codewords with the received word. In that case the number of computations in the worst case equals

$$2^k m\mathrm{PM}(k)+2^k\mathrm{DT}(mk,t).$$

That is

$$B_{\mathrm{exh}}(k,m)=2^{k+1}mk \tag{7}$$

binary operations.

In Table 2 in Appendix A one can see how these complexities compare.

4 Discussion of the results

Roughly speaking we may say that for large k and m,

$$\begin{aligned} B_{\text{all}}(k,m,\lfloor t/m \rfloor) &\approx m^2 k^{\lfloor t/m \rfloor +1}, \\ B_{\text{some}}(k,m,\rho,\lfloor t/\rho \rfloor) &\approx \rho m k^{\lfloor t/\rho \rfloor +1}, \\ B_{\text{exh}}(k,m) &\approx 2^k mk. \end{aligned}$$

So

$$\begin{aligned} B_{\text{exh}}(k,m) : B_{\text{all}}(k,m,\lfloor t/m \rfloor) &\approx 2^k : mk^{\lfloor t/m \rfloor} \\ B_{\text{exh}}(k,m) : B_{\text{some}}(k,m,\rho,\lfloor t/\rho \rfloor) &\approx 2^k : \rho k^{\lfloor t/\rho \rfloor} \end{aligned}$$

If $\rho < m$, the exponent of k is larger in $B_{\text{some}}(k,m,\rho,\lfloor t/\rho \rfloor)$ than in $B_{\text{all}}(k,m,\lfloor t/m \rfloor)$. So, not surprisingly we observe that the complexity of decoding a code with all circulants of full rank is less than decoding a code with not all circulants of full rank.
Furthermore, if k is fixed then for increasing m the difference between the complexities $B_{\text{all}}(k,m,\lfloor t/m \rfloor)$ and $B_{\text{exh}}(k,m)$ becomes less significant. Computer evidence shows that $\lfloor t/m \rfloor$ remains small.
On the other hand, if m is fixed and we increase k, then the complexities $B_{\text{all}}(k,m,\lfloor t/m \rfloor)$ and $B_{\text{exh}}(k,m)$ lie further apart.
The QC-codes in Table 2 in Appendix A, confirm the above analysis. Furthermore, for those QC-codes the decoding algorithm presented here is always better, and often much better, with respect to the complexity, than the decoding algorithm that compares the received word with all the codewords.

References

[1] Ronald W. Bulterman, "Cyclic representatives in lexicographic order", private communication.

[2] F.J. MacWilliams and N.J.A. Sloane, "The Theory of Error-Correcting Codes", North Holland Publishing Company 1977, ISBN 0 444 85009 0.

[3] J.H. van Lint and J.W. Nienhuys, "Discrete Wiskunde", Academic Service, Schoonhoven, ISBN 90-6233-368-0.

[4] Henk van Tilborg, "On quasi-cyclic codes with rate $1/m$", IEEE Trans. On Inform. Theory, Vol. IT-24, No.5, Sept. 1978.

A Quasi-cyclic codes with circulants of full rank

In Table 2 we list, for a certain range of parameters, QC codes that have the additional property that a large number of circulant-matrices G_i, $0 \le i \le m-1$, has full rank, i.e. the generator-polynomials $g_i(x)$ that correspond with these circulant-matrices satisfy $\gcd(g_i(x), x^k - 1) = 1$. The generator-polynomials are listed in octal form. For example $23_o = 010\ 011_b \rightarrow x^4 + x + 1$. The values between brackets, are the generator-polynomials that correspond with a circulant-matrix that is not of full rank.
In column 6 we give the number of binary operations B_{syn} that Algorithm 1 or 3 needs in the worst case to decode a received word (see equations (5) and (6)). It depends on the number of circulants in the generator-matrix that are of full rank, whether B_{syn} equals B_{all} or B_{some}. In column 7, we give the number of binary operations B_{exh} that is needed in the worst case to decode a received word by just comparing it with all the codewords (see equation (7)). In column 8 we compare those two complexities by dividing the second one by the first one.
In some cases we can find a QC code that has minimum distance d one less than the best known minimum distance for those parameters, but that can correct the same number of errors (thus d is odd) and furthermore that has the property that all the circulants have full rank, while the QC codes with the best known minimum distance have not. These codes are also listed in the table, because

their decoding complexity is of interest. Their minimum distance is marked with an asterix.
To find QC codes with $k = 5$ until $k = 11$ and $k = 12, m = 2, 3$ and $k = 13, m = 2$ we performed an exhaustive search, so QC codes with the same parameters, that have more circulants of full rank do not exist.

k	m	d	$\tau = \lfloor \frac{t}{\rho} \rfloor$	$g_0(x), \ldots, g_{m-1}(x)$	B_{syn}	B_{exh}	$B_{\text{exh}}/B_{\text{syn}}$
5	2	4	0	1, 7	15	650	42.7
5	3	7	1	1, 7, 13	267	960	3.6
6	2	4	0	1, 13	17	1536	90.4
6	3	8	3	1, (7), (27)	1481	2304	1.6
6	4	10	1	1, (7), 13, 15	761	3072	4.0
6	5	14	1	1, (7), 13, 15, 37	1278	3840	3.0
7	2	4	0	1, 7	19	3584	188.6
7	3	8	1	1, 23, 37	451	5376	11.9
7	4	12	1	1, 7, 25, 37	886	7168	8.1
7	5	16	1	1, (5), 23, 37, 57	1628	8960	5.5
8	2	5	2	1, (27)	872	8192	9.4
8	3	8	1	1, 15, 57	561	12288	21.9
8	4	12	1	1, 7, 25, 37	1104	16384	14.8
8	5	16	1	1, 15, 57, 67, 73	1828	20480	11.2
9	2	6	2	1, (117)	1170	18432	15.8
9	3	9*	1	1, 37, 45	683	27648	40.5
9	3	10	2	1, (53), 67	3234	27648	8.5
9	4	14	2	1, (43), 57, 73	6344	36864	5.8
9	5	18	2	1, (27), 37, 45, 57	10500	46080	4.4
10	2	6	1	1, 57	287	40960	142.7
10	3	10	1	1, 57, 67	817	61440	75.2
10	4	16	1	1, 67, 127, 163	1612	81920	50.8
10	5	19*	1	1, 43, 75, 117, 127	2230	46080	20.7
10	5	20	2	1, 57, 67, 73, (113)	13912	102400	7.4
11	2	7	3	1, (267)	7064	90112	12.8
11	3	11	1	1, 57, 133	963	135168	140.4
11	4	16	1	1, 37, 127, 133	1902	180224	94.8
11	5	21	2	1, 75, 127, 155, 235	17689	225280	12.7

Table 2, *first part*

k	m	d	$\tau = \lfloor \frac{t}{\rho} \rfloor$	$g_0(x), \ldots, g_{m-1}(x)$	B_{syn}	B_{exh}	$B_{\text{exh}}/B_{\text{syn}}$
12	2	8	1	1, 573	391	196608	502.8
12	3	11*	1	1, 57, 133	1121	294912	263, 1
12	3	12	2	1, (175), 233	7051	294912	41.8
12	4	17	2	1, (137), 163, 255	13788	393216	28.5
12	5	24	3	1, 73, (373), 625, (1255)	68775	491520	7.1
13	2	7	3	1, (653)	13158	425984	32.4
13	3	12	1	1, 277, 313	1291	638976	494.9
13	4	19	3	1, 135, (563), 657	70028	851968	12.2
13	5	25	2	1, 2767, 3753, 3575, 3277	28010	1064960	38.0
14	2	8	1	1, 727	511	917504	1795.5
14	3	13	2	1, 453, 1761	10703	1376256	128.6
14	4	20	2	1, 711, 771, 2273	20882	1835008	87.9
14	5	25*	2	1, 1775, 3623, 4565, 5515	34454	2293760	66.6
14	5	26	3	1, 1353, 1511, 3667, (5775)	153424	2293760	15.0
15	2	8	1	1, 2355	577	1966080	3407.4
15	3	14	2	1, 14777, 27537	12937	2949120	228.0
15	4	20	2	1, 2213, 2275, 2617	25282	3932160	155.5
15	5	28	4	1, (1071), 1751, 13267, (13357)	541079	4915200	9.1
16	2	8	1	1, 557	647	4194304	6482.7
16	3	14	2	1, 16731, 16757	15553	6291456	404.5
16	4	22	2	1, 1505, 10723, 17331	30352	8388608	276.4
16	5	28	3	1, 625, 1167, 12343, (15525)	255528	10485760	41.0

Table 2: QC codes with most circulants of full rank

13. Is Trellis Decoding for Adder Channel Feasible?

Garik Markarian, Bahram Honary
Communications Research Centre, Lancaster University, Lancaster LA1 4YR, UK

1 Introduction

In a multiple access (MA) adder channel, the channel output symbol value is the arithmetic sum of the input symbol values [1,2]. The original model of such a channel was proposed by Kasami and Lin [2] and represents a uniquely decodable code pair of block length $n=2$, as is shown in Table 1.

Table 1. Conventional MA Coding Scheme With 2 Users

User 2\ User 1	(00)	(11)
(00)	(00)	(11)
(01)	(01)	(12)
(10)	(10)	(21)

In this Table, User 1 has only two codewords, $C_1=(00,11)$, and User 2 has three code words, $C_2=(00,01,10)$. Since all the sum codewords are distinct, the decoder can reconstruct the two messages without ambiguity. The overall rate of this MA coding scheme is $R=1.292$ which is better that time sharing [1,2]. An adder channel permits potentially efficient transmission by several users without sub-division in time or frequency. However, practical application of this type of MA is restricted due to the poor synchronisation properties and lack of efficient maximum likelihood decoding techniques [1]. Although a number of different trellis decoding techniques have been proposed [3,4], the general solution of this problem is remaining unsolved due to the non-linear structure of the overall code.

In this paper, we propose a technique that allows trellis design for a binary adder channel. The technique is based on the recently introduced concept of Shannon product of trellises [5] and is applicable to MA coding scheme with M $(M \geq 2)$ users. In this paper, we present a novel decoding technique for the M-choose-T system and we show that a sufficient performance improvement could be achieved without the need for additional information regarding the active users. This approach termed *look-ahead/look-back*, is based on the assumption of finite time transmission by the set of active users and a metric accumulation process. Initially, the system is tested under the assumption of perfect block and bit synchronisation. For the quasi-synchronous system, we use an existing coding scheme from the M-choose-T system and we perform a shifting operation on the user's codewords. The overall trellis structure is a modified version with one sub-trellis corresponding to the synchronised case and a further n sub-trellises were n is the maximum number of possible bit shifts. The modified decoding procedure for the quasi-synchronous case shows a further improvement when compared to the conventional hard decoding techniques.

2 Trellis Design for Binary Adder Channel with M-users

Consider the *M-user* binary adder channel, where *i-th* user is employing (n,k_i,d_i) code C_i, $i=1,2,...M$. We define an overall code $C=\sum_{i=1}^{M} C$ as the set of all possible sums $c=\sum_{i=1}^{M} c$, where c_i is a codeword from C_i, $c_i \in C_i$, and addition represents algebraic sum.

Let T_i, $i=1,2,...M$, be a *BCJR* trellis of C_i [6,7,8], with the following state and branch profiles [5]:

$$N^i = [\, N^i_0, N^i_1, ..., N^i_n] \qquad (1)$$
$$B^i = [\, B^i_1, B^i_2, ..., B^i_n]$$

where the N_t^i, $t=0,1,2,...n$, $i=1,2,...M$, represent the number of states at *t-th* level of T_i, and the B_l^i, $l=1,2,...n$, represent the number of branches at the *l-th* depth of T_i. It is known that such a trellis can be obtained for any linear block code [6,7,8] and although for non-linear block codes the general solution is still unknown, it is always possible to design a "trivial" trellis with parallel paths. Let also two states $S_j^{(t-1)}$ and $S_m^{(t)}$ of the adjacent levels $(t-1)$ and t of T_i be connected by one (or more) directed branch $\omega_t^i\, [S_j^{(t-1)}, S_m^{(t)}\,]$, labelled as x_t^i/c_t^i, where x_t^i and c_t^i represent input and output symbols in C_i.

The overall trellis, *T*, of *C* can be obtained as a Shannon product [5,13] of component trellises $T=T_1 * T_2 * ... * T_M$, with the following state and branch profiles:

$$N(t) = [\, \prod_{i=1}^{M} N^i_0, \prod_{i=1}^{M} N^i_1, ..., \prod_{i=1}^{M} N^i_n] \qquad (2)$$
$$B(t) = [\, \prod_{i=1}^{M} B^i_1, ..., \prod_{i=1}^{M} B^i_n]$$

In this trellis two states of adjacent levels are connected by a branch

$$\omega_t = \omega^1_t[S^{(t-1)}_j, S^{(t)}_m] + \omega^2_t[S^{(t-1)}_o, S^{(t)}_q] + ... + \omega^M_t[S^{(t-1)}_x, S^{(t)}_y] \qquad (3)$$
$$j,o,...,x = 1,2,...,N_{t-1}$$
$$m,q,...,y = 1,2,...,N_t$$

if and only if the trellis T_i has a branch $\omega_t^i\, [S_j^{(t-1)}, S_m^{(t)}\,]$ (addition in (3) represents algebraic sum). At the *t-th* depth of the trellis, branches are labelled as $x^1{}_t, x^2{}_t, ... x^M{}_t/c_t$, where the $x^i{}_t$ $i=1,2,...M$, $t=1,2,...\, n$, represent the *t-th* information symbol in the *i-th* code C_i, and the c_t $t=1,2,...n$, represent the *t-th* encoded symbol in the overall code, *C*.

<u>*Proposition 1:*</u> The designed trellis, *T*, is a *syndrome (BCJR)* trellis of *C*.
<u>*Proof:*</u> see [5].

The technique is best illustrated by the means of the following examples.
Example 1. Let our aim be to obtain a trellis structure for the 2-user coding scheme introduced in [9]. In this scheme, both users employ block codes of length $n=4$:

$C_1=(0000,0011,1100,1111)$
$C_2=(0000,0001,0010,0100,0101,0110,1000,1001,1010)$

and information rate of the overall code is equal $R=1.292$. Code C_1 is a linear *(4,2,2)* code ($n=4$, $k_1=2$, $d_1=2$) with the generator matrix given as follows:

$$G_1=\begin{bmatrix}1 & 1 & 0 & 0\\ 0 & 0 & 1 & 1\end{bmatrix}$$

and code C_2 is a binary image of the ternary *(2,2,1)* code with the following symbol mapping:

$$"0"\Leftrightarrow (00);$$
$$"1"\Leftrightarrow (01);$$
$$"2"\Leftrightarrow (10);$$

Trellises, T_1 and T_2, of these codes can be designed using any of the known techniques (e.g. [6,7,8]) and are shown in Figure 1a. As it follows from this Figure, the component trellises have the following state and branch profiles:

$$N^1(t)=[1,2,1,2,1]; \qquad N^2(t)=[1,3,1,3,1];$$
$$B^1(t)=[2,2,2,2]; \qquad B^2(t)=[3,3,3,3]$$

Following the procedure outlined above, the overall trellis will have the following state and branch profiles:

$$N\ (t)=[1,6,1,6,1];$$
$$B\ (t)=[6,6,6,6];$$

and is shown is Figure 1b (in this Figure symbol "*" in the *i-th* position of a branch label denotes absence of the corresponding information symbols in the *i-th* component code, C_i)

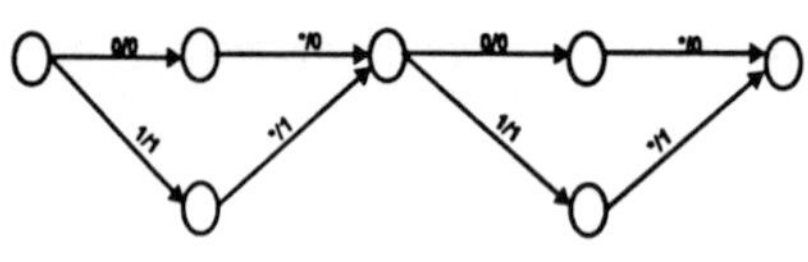

Fig. 1. Trellis Structure of Component code T_1

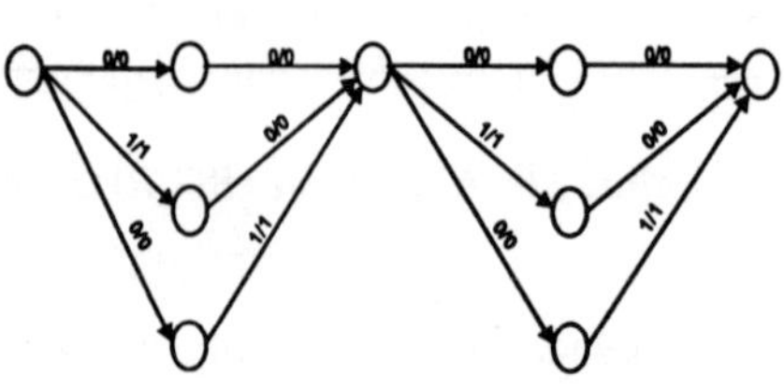

Fig. 2. Trellis Structure of Component code T_2

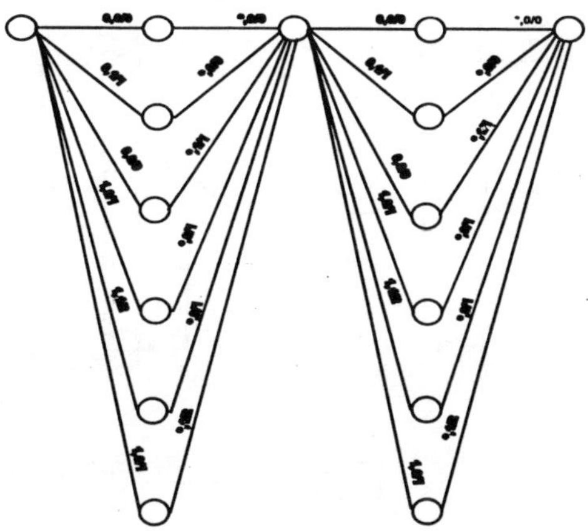

Fig. 3. Trellis Structure of the Overall code

3 Trellis Decoding in the M-choose-T Synchronous Scheme

The *look-ahead/look-back* technique is based on the codes developed in [10]. In this case, the overall coding scheme is represented by a trellis diagram from which the composite and constituent codewords are decoded and the individual user's information vectors extracted. Furthermore, such a technique attempts to identify the set of active users without the need for a special signalling codeword; in fact no extra information is required on the set of active users at all. This approach is based on the assumption that a set of active users will be active during a finite period (frame) in which they will transmit their information. The system keeps track of several received composite codewords and makes a decision when it is confident that a set of users has been active for some time. The amount of time and the number of codewords required to make a sensible decision may be fixed or variable and will be determined upon the system's requirements.

3.1 Decoding Procedure

The *look-ahead/look-back* decoding steps are as follows :

Step 1) Implement *look-ahead* operation by performing maximum likelihood decoding (MLD) on all received composite codewords. At the same time, keep track of metric values accumulated over the transmission period.

Step2) *look-back* to the metric values and make a decision on the set of active users by identifying the decoder with the least accumulated metric.

The technique is illustrated by means of a simple example in which M=3 and T=2. The number of simple decoders required is 3 and the overall decoder is shown in Fig. 4.

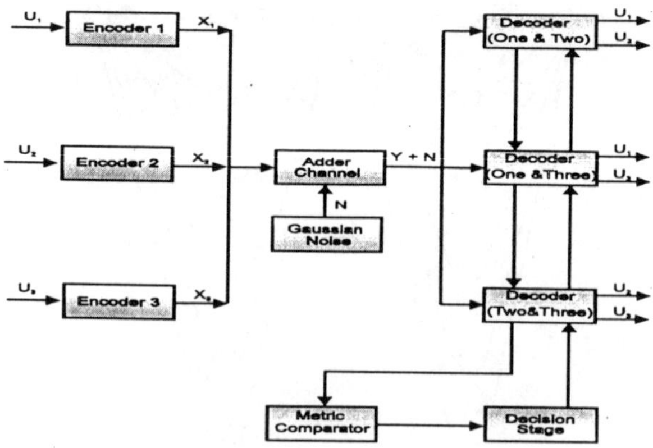

Fig. 4. Block diagram for the M=3-choose-T=2 scheme

In Fig. 4, the system selects a set of users for transmission, since synchronisation is assumed, the decoders know the start and end of each composite codeword block. During transmission, the decoders will attempt to decode the received data according to their structures; and at the end, the identification process and decision making are performed by the *Metric Comparator* and *Decision Stage* blocks respectively.

3.2 Metric Calculation and Accumulation

The overall metric accumulation process is based on calculating the Euclidean distance for each received composite codeword and updating the metric for the next codeword. Each codeword is decoded according to it's own minimum metric, but an overall metric is updated based on the accumulation of all metrics calculated during the transmission period. Since the scheme is implemented in an adder channel, the output symbol with added noise will be :

$$s = \sum_{i=1}^{T} x_i + N \tag{4}$$

where x_i, $i=1, 2, \ldots, T$, is the *ith* user channel input symbol , s the composite channel output symbol and N the Gaussian noise random variable. In each of the decoders in Fig. 4, the Euclidean distance between each received composite vector and all the possible composite codewords is expressed as :

$$\|d_j\|^2 = \sum_{i=1}^{n} \left[r_i - \left(\sum_{i=1}^{T} x_i + N \right) \right]^2 \tag{5}$$

where r_i and n are the received data and the number of bits in the sequence (codeword) respectively and $\|d_j\|$ means that the distance is a metric where $j=1,2,\ldots, {}^{M}C_{T}$ refers to the number of decoders in such a system.

The total metric value accumulated over a transmission period τ can be expressed as follows :

$$\sum_{t=0}^{\tau} \|d_j\|^2 = \sum_{t=0}^{\tau} \sum_{i=1}^{n} \left[r_i - \left(\sum_{i=1}^{T} x_i + N \right) \right]^2 \tag{6}$$

At the end of transmission, the system makes a decision on which set of active users have combined in the channel by identifying the decoder which corresponds to the smallest accumulated metric over τ .

$$\sum_{t=0}^{\tau}\left\|d_j\right\|^2_{\min^k} = \min\left\{\sum_{t=0}^{\tau}\left\|d_1\right\|^2_{\min^1}, \sum_{t=0}^{\tau}\left\|d_2\right\|^2_{\min^2}, \ldots, \sum_{t=0}^{\tau}\left\|d_{^M C_T}\right\|^2_{\min^{^M C_T}}\right\} \tag{7}$$

where $j=k=1,\ 2,\ \ldots,\ {}^M C_T$.

4 Trellis Decoding in the T-user Quasi-Synchronous Scheme

It is a well known fact that reliable decoding in the MA adder channel is based on code constructions that satisfy the unique decodability criteria; this is not the case when the users do not maintain block synchronisation. The search for good coding schemes for the asynchronous MA adder channel is limited as previous research has shown that it is much more difficult to design a certain number of asynchronous codes than to design the same number of synchronous ones. The asynchronous MA adder channel can be reduced to the quasi-synchronous by the following scheme [11], [12] : at the beginning of data transmission, each encoder sends a synchronising sequence to the decoder. The synchronising sequence has the property that, upon reception of a synchronising sequence from one encoder, the decoder can establish block synchronism with it, regardless of what codeword the other encoder is transmitting during the same period. After the reception of synchronising sequences from both encoders, the decoder has established synchronism with each individual encoder, and the asynchronous MA adder channel is reduced to the quasi-synchronous MA adder channel.

Quasi-synchronous uniquely decodable codes (QSUD) have been designed and constructed in [3] based on the definition above. In this paper, we define an alternative version of the quasi-synchronous scheme where the decoder knows the block position of the composite codeword without the need for a synchronisation sequence. We shall use this definition of the quasi-synchronous MA adder channel to investigate the performance of this scheme.

4.1 Description of System

In this case, we use an existing 2-user uniquely decodable code in [10] and we shift one of the user's codeword . The original codes for such a scheme are presented below in Tab. 2.

Tab. 2. 2-User Synchronous Scheme

$User_2$\$User_1$	000	001	110
000	000	001	110
010	010	011	120
101	101	102	211

In this table, $User_1$ and $User_2$ have three codewords as C_1=(000,001,110) and C_2=(000,010,101) and the overall rate is $R=1.057$. If $User_2$ codewords were shifted by 1 bit and assuming the first bit entering the decoder is the leftmost bit, then shifting is performed from right to left and the result for such shifted version is presented in Tab. 3.

Tab. 3. 2-User Quasi-synchronous Scheme

$User_2 \backslash User_1$	000	001	110
00X (000 or 001)	000 001	001 002	110 111
10X (101 or 100)	101 100	102 101	211 210
01X (010 or 011)	010 011	011 012	120 121

Inspection of Tab. 3 reveals that the 2-user scheme is not uniquely decodable which makes the decoding process unreliable and delivering correct data to its corresponding destination will not be achievable in some cases . Although this can be seen as a drawback, we will use this effect for good cause. It can also be seen in Tab. 3 that $User_2$ constituent codes have doubled due to the shifting operation performed and therefore $C_2 = (000,001,101,100,010,011)$ and consequently the overall rate has increased to $R=1.389$.

4.2 Trellis Construction and Representation

The original trellis structures for the $User_1$ and $User_2$ codes in Tab. 2 are illustrated in Fig. 5 and 6 respectively.

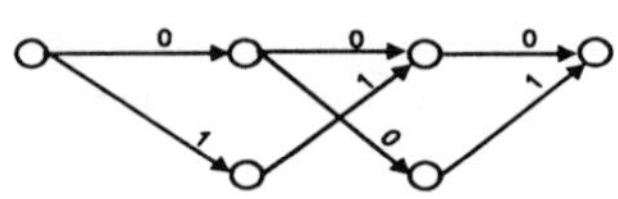

Fig. 5. Trellis Structure of $User_1$ Code

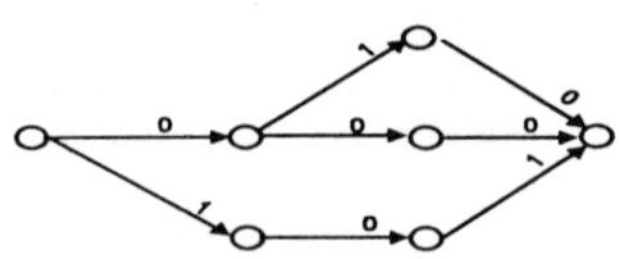

Fig. 6. Trellis Structure of $User_2$ Code

Since $User_2$ constituent codewords have increased from three to six, the trellis diagram in Fig. 6 has to be modified in order to construct the new structure. Tab. 3 can be seen to have three main rows, each having two sub rows : upper and lower. The upper sub-rows can be seen to represent the original synchronous scheme in Tab. 2 and the lower sub-rows represent the remainder from the effect of the shifting operation. Therefore, trellis construction for the $User_2$ case will be based on upper and lower sub-trellises. The modified version for such a code is shown in Fig. 7.

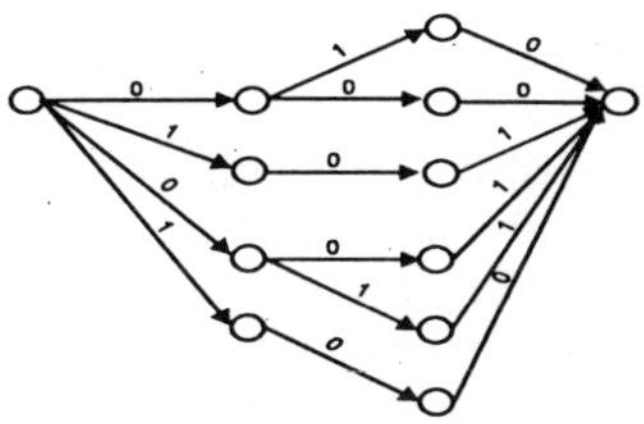

Fig. 7. Modified Trellis Structure of $User_2$ Code

In order to construct the composite trellis structure of this scheme, the Shannon Product of trellises is applied to the codes in Fig. 5 & 7. Since $User_2$ trellis has been modified as illustrated in Fig. 7, the composite trellis should also be made up of two sub-trellises. Consequently, the overall composite trellis is illustrated below in Fig. 8.

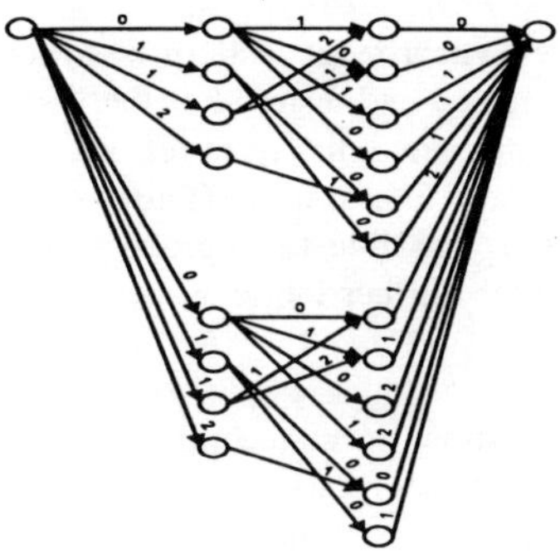

Fig. 8. Shannon Product Trellis Structure of $User_1$ + $User_2$ Code

4.3 Combined Maximum Likelihood Trellis and Erasure Decoding

The decoding procedure used here is termed combined maximum likelihood (ML) trellis and erasure decoding. The term ''Combined'' is employed because the decoder performs ML estimation on the received composite data, but in some cases, the ML process can only decode 2 constituent bits out of 3. The last bit for each individual users is erased if the composite codeword is not uniquely decodable. Therefore, to make the decoding process complete, the users have to either request a retransmission of the last bit or make a half chance correct decision. Error control coding can be used to correct erasures or to correct errors and erasures simultaneously. If a code has minimum distance d_{min} , any pattern of p or fewer erasures can be corrected if

$$d_{min} \geq p + 1 \tag{8}$$

In our case, d_{min}=1 for the individual user's and composite codewords and therefore erasure correction is not possible. Consequently, a bit error in the individual user's codes is inevitable sometimes.

The decoding algorithm for the quasi-synchronous 2-user coding scheme is based on the following :

Step 1) Perform MLD on the received composite codeword.

Step 2) If the received codeword is uniquely decodable, retrieve the individual user's codewords and in the $User_2$ case retrieve the original code.

Step 3) If the received codeword is a duplicate, decode the first 2 bits of the individual user's codes and erase the last bit.

Step 4) To complete the decoding process, request a 1 bit transmission from each user (independently) or retransmit a new composite codeword.

Step 4 is repeated until the composite codeword received is uniquely decodable in order to extract the component codes without ambiguity.

5 Conclusion

The performances of the synchronous and quasi-synchronous MA adder channel schemes have been investigated and tested. For the M-choose-T synchronous case, a sufficient performance improvement has been realised without the need for additional information on the set of active users. In the T-user quasi-synchronous system, it has been demonstrated that it is possible to improve such a scheme by modifying the constituent codes in use with further increase in overall code rate at the expense of decoder complexity.

There are a number of questions in connection with the quasi-synchronous system. Firstly, decoding in the M-choose-T case and the identification process cannot always be implemented reliably in the quasi-synchronous channel. This is due to the fact that the codes used here do not exhibit the same characteristics when a shifting operation takes place and modified trellis structures for the different set of active users cannot be constructed in an M-choose-T quasi-synchronous environment. The search for efficient coding schemes that can be applied to such an environment is a subject of further investigation. One possible solution is to modify the metric calculation and accumulation process by introducing a new metric other than the Euclidean distance. The idea stems from the fact that when the MA adder channel is synchronous, the Euclidean distance metric is used as the noise in the channel is assumed to be Gaussian. This is not the case when the channel is quasi-synchronous.

6. References

1. Farrell P.G. "Survey of channel coding for multi-user systems", in *New Concepts in Multi-User Communication,* Editor: J.K.Skwirzynski, The Netherland, Sijthoff and Noordhoff, 1981, pp.133-159.
2. Kasami T., Lin S. "Coding for a multiple access channel", *IEEE Transactions on Information Theory,* vol.22, No.2, 1976, pp.129-137.
3. Lin S., Wei V. "Non-homogeneous trellis codes for the quasi-synchronous multiple-access binary adder channel with two users", *IEEE Transactions on Information Theory*, vol.32, No.6, 1986, p.p.787-796 .
4. da Rocha V., Markarian G., Honary B. "Trellis decoding of block codes for the 2-user binary adder channel", *Proceedings of the ICC'93*, Geneva, vol.2/3, pp.821-826.
5. Sidorenko V., Markarian G., Honary B. "Code trellises and the Shannon product", *Proceedings of the seventh Joint Swedish-Russian International Workshop on Information Theory*, July 1995, St.-Petersburg, Russia, pp.220-224.
6. Wolf J. "Efficient maximum likelihood decoding of linear block codes using a trellis", *IEEE Transactions on Information Theory*, vol.24, No.1, 1978, pp.76-80.
7. McEliece R. "The Viterbi decoding complexity of linear block codes", *Proceedings of IEEE Information Theory Symposium,* June 1994, Trondheim, Norway, pp341.
8. Honary B., Markarian G. " Trellis decoding of block codes : A practical approach ", *Kluwer Academic Publishers,* 1997.
9. Kasami T., Lin S., Yamamura S. "Further results on coding for a multiple-access channels". *Proceedings of Hungarian Colloquium on Information Theory,* Keszthely, Hungary, 1975, pp.369-391.
10. Mathys P. " A class of codes for T active users out of N for a multiple access communication system, " *IEEE Trans. Inform. Theory,* vol. 36 pp. 1206-1219, Nov. 1990.
11. Kasami T., Lin S., Wei V., Yamamura S. " Graph theoretic approaches to the code construction for the two-user multiple access binary adder channel, " *IEEE Trans. Inform. Theory,* Vol. 29, pp. 114-130, 1983.
12. McEliece R. J., Posner E. C. "Multiple access channels without synchronisation, " *in Conf. Rec. Int. Conf. on Communications,* Vol. 2, pp. 246-248, 1977.
13. Markarian G., Honary B., Benachour P. " Trellis decoding techniques for the binary adder channel with M-users, " *IEE Proc. on Communications,* Vol. 144, 1997.

CHAPTER 4

Signal Processing in Communications Systems

Based on "Efficient Bandwidth Utilisation for CDMA Using Convolutional Codes" by I. Band and D. Cruickshank, which appeared in *Proceedings of the IEEE Global Telecommunications Conference (Globecom),* London, UK, Nov. 1996, Vol.2, pp 1306-1310.

14. CDMA Spread Spectrum Systems using Spreading Sequences Combined with Convolutional Codes

P.M. Grant I.W. Band D.G.M. Cruickshank

Department of Electrical Engineering, University of Edinburgh, EH9 3JL, UK

ABSTRACT This paper investigates the use of non-linear convolutional coding techniques for use in a DS-CDMA cellular communication system. Particular attention is paid to the relative proportions of convolutional coding and direct sequence spreading which offer the most efficient use of the available processing gain. The dependency of the optimal strategy on various factors, including background noise level, number of active users and coding power is discussed. Future developments are indicated including the possible application of this system to multipath channels, and the use of other coding structures.

1 Introduction

The demand for increased capacity of cellular mobile phone systems is projected [1] to reach some 40 million users in Europe alone by the end of the century. Current digital systems such as the European global system for mobiles (GSM), which use time as the discriminant between users to provide multiple access (TDMA), are in competition with potentially higher capacity [2] code division multiple access (CDMA) systems, such as that detailed in the IS-95 interim standard for mobile communications [3].

In this paper, which is derived from [4], we study the downlink (base station to mobile) of a bit and chip-synchronous direct sequence code division multiple access (DS-CDMA) cellular communication system, and consider methods of combatting errors by introducing non-linear convolutional codes. The efficient use of signal bandwidth is of primary concern in practical cellular communications applications, so we consider the performance of various systems comprising different proportions of convolutional encoding and direct sequence spreading, whilst maintaining a given overall bandwidth expansion.

Boudreau [5] has calculated the theoretical performance (in terms of Chernoff upper bounds) of systems employing convolutional coding (and trellis coding) as a natural extension of the conventional CDMA principle. The performance of actual systems, however, may be significantly superior (as much as 1.5 dB in some cases); this improvement being a non-linear function of E_b/N_0. The limited amount of simulations in [5] means that it is instructive to consider the simulated performance of such systems.

The noisy environment considered here is due only to multiple access interference (MAI) from competing users and additive white Gaussian noise (AWGN) which may be taken to represent either system or thermal noise. The performance in AWGN (in terms of the average number of errors over a large number of Monte Carlo simulations) may be expressed in terms of the decibel form of the quantity E_b/N_0, given by

$$\frac{E_b}{N_0} = \left(\frac{G_P}{2\sigma^2}\right) \tag{1}$$

where σ^2 is the signal to noise power variance and G_P is the bandwidth expansion or processing gain used.

The next section describes convolutional encoding and decoding in more detail, while section 3 presents some results. Finally, section 4 discusses the implications of these results with emphasis on future developments.

2 Coding Strategies

In this section, we briefly describe the use of non-linear convolutional codes to decrease the probability of error, discuss the decoding algorithm employed and give details of the particular configurations used in this study.

Convolutional encoding[6] is a popular form of forward error correction (FEC), in which the output C data chips are a function of the incoming b data bits and the connections of a set of serial shift registers to a set of modulo-2 adders. The number of shift registers is termed the constraint length, K, and the rate R of the code is defined as the ratio of input to output digits (b/C). In this study, we confine simulations to convolutional codes with $b = 1$ and values of C of 2,4,8 and 16.

The Viterbi algorithm[7] is a maximum likelihood (ML) decoding algorithm [8], which proceeds by constructing a trellis; the surviving branches of which are determined by a comparison between the actual received and possible received signals. If the incoming data samples are first quantised to binary data the process is termed hard decision, while if the floating point values are retained for the comparison, the decoder is said to be employing soft decision. The latter improves performance by around 2 dB for most values of E_b/N_0 [9], and is the scheme adopted here.

After an appropriate amount (corresponding to the survivor path length, L) of data has been received and analysed, the algorithm makes a decision on the optimal path back through the trellis and hence deduces the initial data bit.

Thus, a decision is only made on a particular data bit after code sequences from L data bits have been received.

Convolutional Code Configurations: It has been shown [9] that, provided the survivor path length is chosen greater than 4-5 times the constraint length, the performance of the Viterbi algorithm is not adversely affected. In the investigations here, L is set to 32, and the values for R and M are shown in Table 1, which provides the octal format of the tap connections used for the convolutional codes. It may be seen that the rates of the codes and the lengths

Table 1: Convolutional codes used in the simulations

K	$R = \frac{1}{2}$ $M = 255$ $G_P = 510$	$R = \frac{1}{4}$ $M = 127$ $G_P = 508$	$R = \frac{1}{8}$ $M = 63$ $G_P = 504$	$R = \frac{1}{16}$ $M = 31$ $G_P = 496$
3	(5,7)	(5,7,7,7)	(7,7,7,5,5,5,7,7)	(7,7,7,5,5,5,7,7, 7,7,7,5,5,5,7,7)
4	(17,15)	(13,15,15,17)	(17,17,13,13,13,15,15,17)	(17,17,13,13,13,15,15,17, 17,17,13,13,13,15,15,17)
5	(23,35)	(25,27,33,37)	(37,33,25,25,35,33,27,37)	(37,33,25,25,35,33,27,37, 37,33,25,25,35,33,27,37)
6	(53,75)	(53,67,71,75)	(53,67,71,75,53,67,71,75)	(53,67,71,75,53,67,71,75, 53,67,71,75,53,67,71,75)

of the corresponding PN-sequences have been selected to maintain an overall processing gain (denoted G_P in the table) of around 500. The choice of tap connections for rates 1/2 ($M = 255$ [1]) and 1/8 ($M = 63$) mirror those in [5] for constraint lengths 3,4 and 5 whilst those for $R = 1/4$ are motivated by the results of Larsen [11]. While it is acknowledged that the repetition of the rate 1/8 code to form the rate 1/16 code may be sub-optimal, especially for the larger constraint-length systems, this configuration is liable to have similar properties to any optimal arrangement, and is thus employed here.

3 Results

In this section, we describe the simulations carried out and present the results of the various scenarios outlined previously. The basic system diagram is shown in figure 1, which also indicates the length of the signal at each of the relevant points.

Referring in more detail to figure 1, the data vector $\underline{x}_1$ of L bits is input to the convolutional encoder, and the resulting LC chips of data are then further spread by the user's pseudo-noise (PN) code to produce a signal of LCM chips. This process is repeated over all active users (with the same convolutional

[1] Strictly, Gold codes of length 255 do not exist[10], however "Gold-like" sequences may be constructed with the required cross-correlation of -1 at the synchronous point. Those used here are obtained from the m-sequences $[8,2,3,4]_s$ and $[8,3,5,6]_s$

code configuration for each user [2]) and the final transmitted signal vector $\underline{s}$ is then constructed by coherently summing the spread signals over all the users. Additive white Gaussian noise, represented by the vector $\underline{n}$ is then added chip by chip to this signal to form the input to the receiver, $\underline{y}$. This noise-corrupted signal is then de-modulated by first correlating with the first user's spreading code, and then passing on the resulting LC encoded data bits to the Viterbi decoder; the output from which forms the estimate, $\underline{\hat{x}}_1$, of the original data vector.

The ratio of the number of errors to the total number of transmitted data values forms the main output statistic (P_e), which may be evaluated as a function either of background noise level, considered in the first four case-studies, or of number of active users for a fixed value of E_b/N_0, considered in the last two.

In the first four cases, Gold codes are used as the PN sequences and there are 20 active users present, while in cases 5 and 6, random codes have been chosen to spread the convolutionally-encoded data, and the values of E_b/N_0 are held fixed at 2.0 dB and 4.0 dB respectively. The choice of random codes is motivated by the fact that in a multipath environment such as those defined in the COST 207 study [12], the presence of intersymbol interference will corrupt the near-orthogonal qualities of Gold codes. Note that in this case, the percentage loading is calculated from the number of active users and the overall processing gain (G_P), rather than the length of the appropriate PN code.

Case 1: $K = 3$**:** The first case considered here is for constraint length 3 and is shown in figure 2. Note that, in the figures, the performance of each system is compared with binary phase shift keying (BPSK) and a conventional DS-CDMA system using 511-chip Gold spreading codes.

It is apparent that in this case, the rate 1/2 system performs better than the other configurations, with the rate 1/8 system also performing well. The relatively poor performance of the rate 1/16 system is probably due to the increased cross-correlation of these Gold codes (20 users with 31-chip spreading means that the multiple access interference is more significant than the longer PN-code systems), combined with the possible sub-optimality of the convolutional codes. The reason for the anomalously poor performance of the rate 1/4 system is less clear, although this may simply be an unfortunate combination of rate and constraint length. Further investigations may provide some insight about this behaviour.

Notwithstanding these comments, even the poorest performing convolutional code is still significantly better than the 20-user conventional CDMA system

[2] This means that the separation of users is achieved solely by the spread spectrum module. A possible development of this which may be considered in the future would be to assign different tap connections for each user

for values of E_b/N_0 greater than about 1 dB.

Case 2: $K = 4$: The results for constraint length 4 systems are presented in figure 3, from which it may be seen that the rate 1/4 and 1/8 systems, which are now very similar, are slightly better than the rate 1/2 system, while the rate 1/16 codes are still suffering from the increased MAI of the 31-chip Gold codes in a similar way to the K=3 case. The improved performance of the rate 1/4 system is probably due to the more diverse use of tap connections here than in the previous case. Where a greater variety of tap connections is available, it is prudent to make as full use of them as possible, thus increasing the power of the code and permitting more resources (bandwidth) to be invested in the direct spreading part, to allow greater capacity.

Thus, while the choice of rate and constraint length are notionally independent, the use of appropriate rate codes, which make use of the full range of available connections for a given constraint length can lead to marked improvements in performance.

Case 3: $K = 5$: Figure 4 shows the results obtained using systems with constraint length 5. This appears to confirm the trend noted in case 2, namely that the systems with rates 1/2 and 1/16 are inferior to those with rates 1/4 and 1/8, which make more efficient use of the increased performance, by virtue of their larger constraint lengths, of the convolutional codes. A point of interest is that now the rate 1/16 codes are better than the rate 1/2 system. This is again due to more efficient use of available connections, which is beginning to compensate for the greater MAI of the 31-chip spreading module, in contrast with the first two cases with their shorter constraint lengths.

The performance of the rate 1/2 system at E_b/N_0 values lower than around 1 dB is also worthy of note, since it is now inferior to conventional CDMA. This is probably also due to the increased power of the convolutional codes, since more errors are introduced from the Viterbi algorithm losing track of the correct path through the trellis. Thus it is not always beneficial simply to increase the constraint length independently, since more errors may be introduced under increased-noise environments. This effect will be examined in more detail in case 5, in which the additional noise will be due to increasing numbers of users.

Case 4: $K = 6$: For constraint length 6, shown in figure 5, the trend outlined previously is reinforced, with the rate 1/8 and rate 1/4 systems better than the rate 1/16 and 1/2 systems, respectively. The rate 1/16 system has better performance than before, since again, more efficient use is made of the available tap connections.

As in case 3, the rate 1/2 system performs significantly poorer than the conventional case for E_b/N_0 values below around 1 dB. This is again due to the Viterbi algorithm producing a higher number of errors when an incorrect path through the trellis is selected. Thus, care needs to be exercised when employing a long constraint length, high rate convolutional code under conditions of high noise.

Case 5: E_b/N_0 **=2.0 dB:** The performance of these systems with varying numbers of active users may be judged from figure 6, for which the background noise level is set at 2.0 dB (equivalent to 5.0 dB signal-to-noise ratio). In this case, at least 1000 errors have been detected for each point on the curves and constraint lengths of 3 and 6 are employed. Random codes are used as the PN sequences to more closely model practical situations, and this means that the trends seen through the previous cases will be accentuated.

For very low loading (around 2%), the systems behave much as would be expected, although the $K = 3$, rate 1/16 system performs even worse than conventional CDMA, and the $K = 3$ rate 1/4 system is only marginally better. In general, the $K = 6$ systems have better performance, due to the increased power of the codes, however it is interesting that even for very low numbers of users, the rate 1/2 $K = 3$ and $K = 6$ systems are very similar.

As the loading is increased, various effects may be seen. In particular, the only systems which continue to outperform conventional CDMA are the $K = 6$ rate 1/4 and $K = 3$ rate 1/2 configurations. These systems represent the most efficient use of the combination of rate, constraint length (and hence code performance) and spreading capacity, and thus have superior performance.

Significantly, the $K = 6$ rate 1/8 combination, which appeared to be the best for the previous cases with 20 users using Gold codes, does not have such good performance using random codes as the number of users is increased. In addition, the $K = 6$ rate 1/2 combination has poorer performance than the equivalent $K = 3$ system. Although this may at first appear counter-intuitive, the reason is due to the increased overall noise causing the Viterbi algorithm to lock on to the wrong path through the trellis, and hence produce more errors, analogous to the noisy situations in cases 3 and 4.

Case 6: E_b/N_0 **=4.0 dB:** The equivalent results for $E_b/N_0 = 4.0$ dB are shown in figure 7. This confirms some of the results in the previous case; specifically, that although the $K = 6$ rate 1/2 system is better for low loading, as the loading is increased its performance becomes worse than the same rate $K = 3$ system, and that the beneficial effects of most convolutional coding systems are lost as the loading increases. As in the previous case, the optimal system as the number of users is increased is that using $K = 6$ rate 1/4 convolutional codes, with 127-chip spreading codes.

4 Discussion & Future Work

This paper has compared the simulated performance of a number of strategies utilising non-linear convolutional codes combined with pseudo-noise spreading codes for use in a processing-gain limited DS-CDMA cellular communication system. Convolutional codes with constraint lengths up to $K = 6$ have been considered, with rates of 1/2, 1/4, 1/8 and 1/16, together with appropriate length spreading sequences.

With an overall processing gain of approximately 27 dB and 20 active users, the optimal strategy seems to be the combination of a constraint length 6, rate 1/8 convolutional code and 63-chip Gold spreading codes, although increasing the constraint length may cause a different rate to be optimal. Increasing the number of users and employing random spreading sequences means that this arrangement is no longer optimal, and that convolutional codes of the same constraint length but rate 1/4 should be used instead. The ill-effects of simply increasing the constraint length without a corresponding adjustment of the code rate are also demonstrated.

These observations serve to exhibit the fine balance that exists between the power of the convolutional code (obtained from its rate and constraint length), the capacity of the spreading sequences and the noise level, whether from the background or from other subscribers on the system.

A development of the work would be to investigate the performance of these techniques in multipath environments such as those defined in the COST 207 study. Alternative methods of convolutional encoding [13] may also prove useful in future studies.

Acknowledgement The authors would like to acknowledge the financial support of this project by the Engineering and Physical Sciences Research Council, through Grant GR/J46401.

References

[1] I. S. Groves, "Advances in Personal Communication Services," *Proceedings of the IEEE 5th International Symposium on Personal, Indoor and Mobile Radio Communications (PIMRC), The Hague, The Netherlands*, vol. 1, pp. 11–14, Sep 1994.

[2] K. S. Gilhausen, I. M. Jacobs, R. Padovani, and L. A. Weaver, Jr., "Increased Capacity using CDMA for Mobile Satellite Communication," *IEEE Journal on Selected Areas in Communications*, vol. 8, pp. 503–514, May 1990.

[3] TIA/EIA/IS-95, *Mobile Station-Base Station Compatibility Standard for Dual Mode Wide-Band Spread Spectrum Cellular Systems.* Telecommunication Industry Association, 1993.

[4] I. Band and D. Cruickshank, "Efficient Bandwidth Utilisation for CDMA Using Convolutional Codes," *Proceedings of the IEEE Global Telecommunications Conference (Globecom), London, UK*, Nov 1996.

[5] G. Boudreau, D. Falconer, and S. Mahmoud, "A Comparison of Trellis Coded Versus Convolutionally Coded Spread Spectrum Multiple Access Systems," *IEEE Journal on Selected Areas in Communications*, vol. 8, pp. 628–640, May 1990.

[6] P. M. Grant, C. F. N. Cowan, B. Mulgrew, and J. H. Dripps, *Analogue and Digital Signal Processing and Coding.* Bromley, UK: Chartwell-Bratt, 1989.

[7] A. Viterbi, "Error Bounds for Convolutional Codes and an Asymptotically Optimum Decoding Algorithm," *IEEE Transactions on Information Theory*, vol. IT-13, pp. 260–269, April 1967.

[8] H.-L. Lou, "Implementing the Viterbi Algorithm," *IEEE Signal Processing Magazine*, pp. 42–52, September 1995.

[9] J. Heller and I. Jacobs, "Viterbi Decoding for Satellite and Space Communication," *IEEE Transactions on Communications*, vol. COM-19, no. 5, pp. 835–848, 1971.

[10] D. V. Sarwate and M. B. Pursley, "Cross-correlation Properties of Pseudorandom and Related Sequences," *Proceedings of the IEEE*, vol. 68, pp. 593–619, May 1980.

[11] K. Larsen, "Short Convolutional Codes with Maximal Free Distance for Rates 1/2,1/3, and 1/4," *IEEE Transactions on Information Theory*, vol. IT-18, pp. 371–372, 1973.

[12] M. Failli, ed., *Digital Land Mobile Radio Communications - COST 207 : Final Report.* Luxembourg: Commision of the European Communities, 1989.

[13] R. F. Ormondroyd and J. J. Maxey, "Performance of Low-Rate Orthogonal Convolutional Codes in DS-CDMA Applications," *IEEE Transactions on Vehicular Technology*, pp. 320–328, May 1997.

Figures

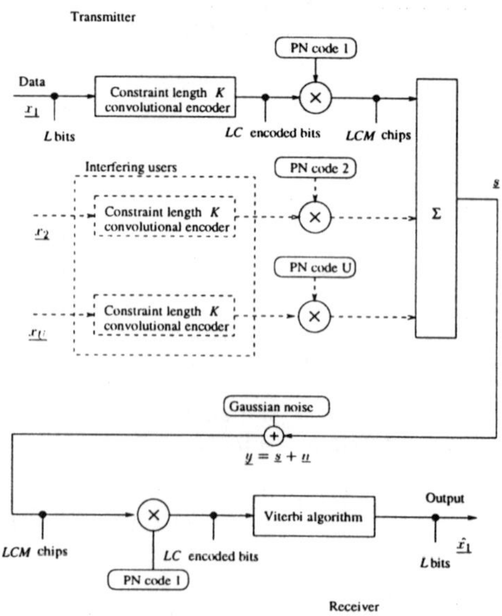

Figure 1: *Flow of data*

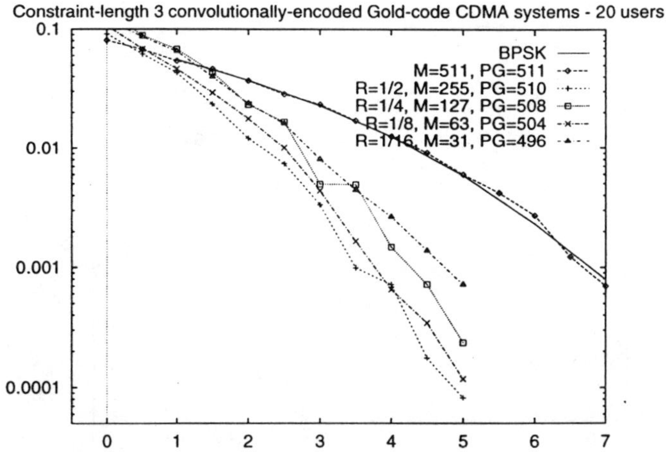

Figure 2: *Performance of K=3 convolutional codes*

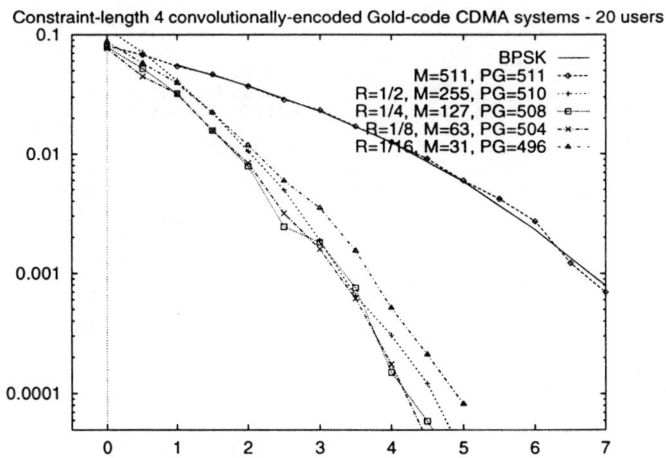

Figure 3: *Performance of K=4 convolutional codes*

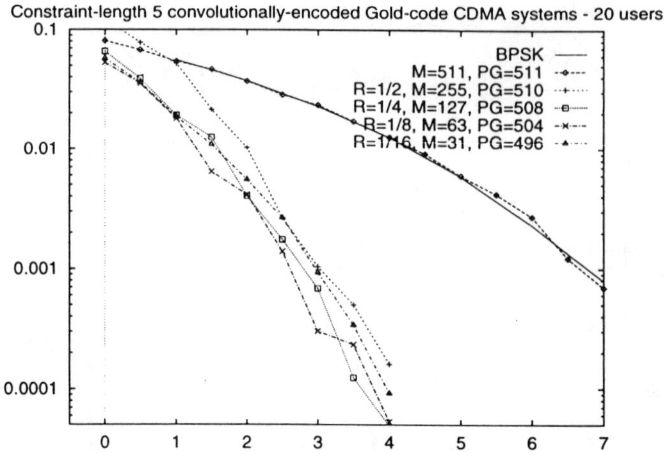

Figure 4: *Performance of K=5 convolutional codes*

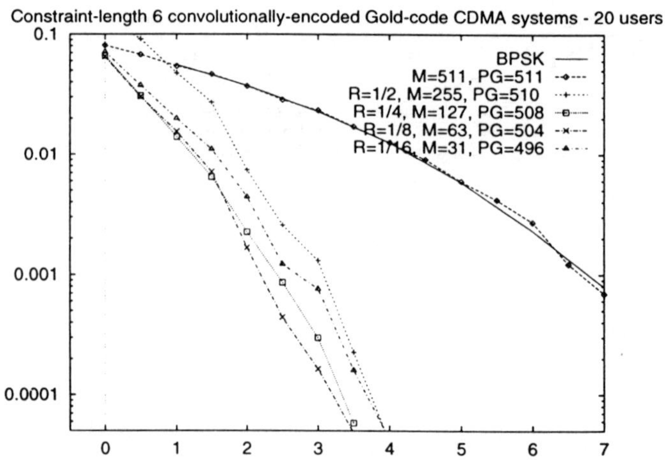

Figure 5: *Performance of K=6 convolutional codes*

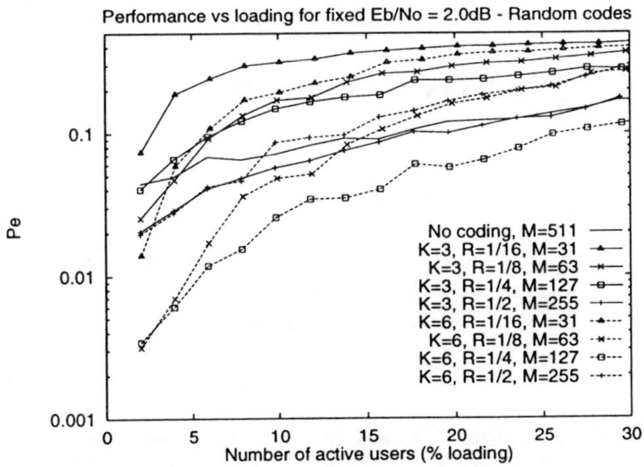

Figure 6: *Performance of K=3 and K=6 convolutional codes for* $E_b/N_0 = 2.0$ *dB*

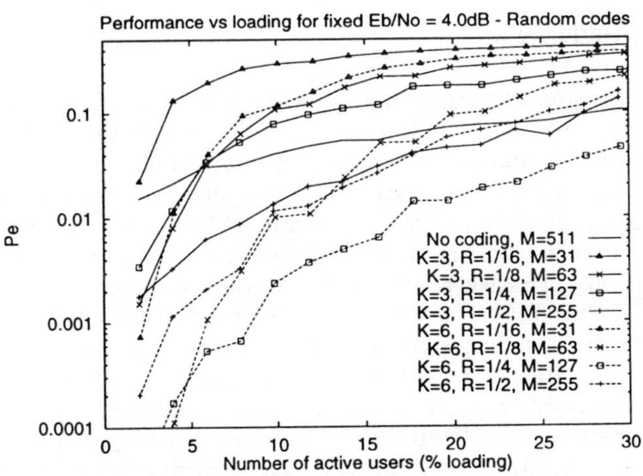

Figure 7: *Performance of K=3 and K=6 convolutional codes for* $E_b/N_0 = 4.0$ *dB*

PGF B/Celebration Conference, Lancaster University, 25-28 January 1998

15. Detection and Classification of DS/SS Codes using Higher-Order Statistics

Peter C J Hill & Ernest R Adams

Communications Electronics & Signal Processing Group
Department of Informatics and Simulation
Cranfield University RMCS

Abstract

Modern advances in digital signal processing are leading to novel intercept receivers which specifically take account of the structural properties of signal waveforms and spectra from smart emitters. In particular higher-order statistical processing provides a means for detecting, and in some cases, also characterising direct-sequence spread spectrum signals. Triple correlation analysis has been used to detect chip-code sequence polynomial generators in full and partial coded windows. The correlation patterns remain discernible for data modulation in noisy channels. Moreover, suitable multivariate analysis allows individual m-sequences to be discriminated in the presence of other co-channel sequences. Blind search techniques for exploring triple correlation patterns of multiple sequences are proposed.

1. INTRODUCTION

Military and civil radio authorities have an interest in detecting and characterising communication signals. The problem now is that both legitimate and unauthorised sources are starting to use low probability-of-intercept (LPI) codec designs such as spread spectrum for terrestrial and satellite radio systems. The main threat is from frequency hopping (FH/SS) and direct sequence (DS/SS) coding where the detection capability of intercept receivers is severely stretched in multiple access (CDMA) scenarios. The low noiselike power spectral density (PSD) of DS/SS presents a particularly difficult detection problem; moreover, blind signal separation in CDMA systems, where the code sequences are not available, is virtually impossible with current technology. High noise, interference and jamming environments considerably exacerbate electronic intercept problems in detecting, classifying, locating, identifying and exploiting radio signals. Fortunately advances in digital signal processing can be used to exploit modern LPI waveforms [1]. Our research at RMCS using statistical processing[†] is providing particularly useful techniques in this respect [2].

2. INTERCEPT OF LPI SIGNALS

The detection of FH/SS is more advanced than that for DS/SS because conventional transmissions are used in each hop-slot and there is now predominant availability of FH radio equipment. Conventional linear receivers employing fast digital processing can sort out multiple FH signals into frequency and direction-of-arrival (DOA) bins by

---

[†] This work has been supported in part by US Air Force,Wright Laboratory, F61708 95 C008

using collateral DF equipment. Application to DS/SS is limited because of the sub-unity signal/noise ratio (SNR) across a very low PSD channel.

Non-linear receivers employ signal processing techniques which effectively produce spectral region(s) where the SNR is large enough to provide limited feature detection i.e. parameters such as centre frequency and chip rate in DS/SS [3]. There are detection limitations with current digital non linear receivers such as single and dual channel delay correlators. The threats for the interceptor include transmission power control, fast coherent hopping FH/SS and QPSK or SQPSK modulation in DS/SSMA. Moreover our research in chip code design for enhancing LPI in DS/SS [4] has shown that analogue spreading sequences based on pseudo random phase samples across a white spectral domain can actually suppress harmonic output from a single channel correlator, as well as providing an adaptive antijamming capability; various statistical chip code modulations are also effective in this respect.

In general DS/SS signals cannot be decoded without knowledge of the transmitted chip codes. For the special case of m-length PN sequences operating in good SNR channels with full code switching, the Massey cryptanalysis technique can be used to decode the linear feedback shift-register (LFSR) generator. The method however is unworkable with partial and/or non-PN sequences and practical DS/SS channels. The decoding task is impossible for DS/SSMA unless code structures are available where it is possible to successively despread CDMA signals using some form of iterative processing. Emitter location is also difficult with DS/SS as chip-coded reference loops are required for DOA estimation using available digital beamformer techniques.

3. STATISTICAL PROCESSING TECHNIQUES

Traditional signal processing in electronic intercept relies on time and frequency analysis via the DFT/FFT algorithm. For nonstationary signals we have to use a short-time FT (STFT) quadratic process such as the Wigner distribution or its smoothed variant (Wigner-Ville distribution) which can theoretically deal with multiple signals such as CDMA without unacceptable cross-terms appearing. Another approach is to use the Wavelet transform as a constant Q-factor STFT. But for digital spread spectrum signals we should somehow take account of the statistical cyclostationarity (periodically stationary) property of the waveforms involved. A simple sideband correlator based on this property has been researched at RMCS [5].

The motivation for our research into DS/SS and CDMA detection using higher-order moments (HOM) and polyspectra (HOS) is that the cross-cumulants of non-gaussian cyclostationary processes are ideally zero and the process is also robust against background noise; moreover, the phase character of signals is recoverable [6]. We have concentrated on triple correlation and bispectrum processing where the two-dimensional pseudo images can characterise the intercepted DS/SS signals through the placement of multiple correlation peaks; simple thresholding in noise retains the essential chip code properties. In fact discrete polynomial theory allows us to decode linear sequences such that, under certain conditions, it is even possible to estimate the LFSR chip code generator(s).

4. TRIPLE CORRELATION OF DS/SS M-SEQUENCES

The discrete triple correlation function (*tcf*) of a complete *L*-length m-sequence $v(i)$ may be evaluated as:

$$C(p,q) = \frac{1}{L}\sum_{i=1}^{L} v(i)v(i+p)v(i+q)$$

Note that the *tcf* is a function of two shifts p and q, and $v(i)$ is assumed to be *L*-periodic. Peaks arise in $C(p,q)$ as, for each p', there is a unique q' for which $v(i+p')v(i+q')=v(i)$, when $C(p',q')=1$. For other $q \neq q'$, $v(i+p')v(i+q)=v(i+r)$ for some shift $r \neq 0$, giving $C(p',q)=-1/L$ from the autocorrelation properties of m-sequences. Figure 1 shows the *tcf* plot of a typical m-sequence and the *tcf* of the same code in a channel SNR of 6 dB is given in Figure 2; the peaks are clearly discernible.

The *tcf* of two superimposed m-sequences consists of the two individual sets of peaks of unit height with additional peaks depending on cross-correlation properties of the sequences [7]. For $L=31$, the magnitudes of cross-correlation peaks lie between 1/31 and 11/31. The uniqueness of the principal peaks for each m-sequence thus leads to a simple test for a particular sequence that is *independent* of the presence of other sequences. The *tcf* of two superimposed preferred sequences is shown in Figure 3.

In practice, data sample lengths will be shorter than the unknown lengths of complete sequences present. If only a partial sequence of length $N(<L)$ is sampled, periodicity may not be assumed in calculating its triple correlation. A partial *tcf* may be estimated:

$$C'(p,q) = \frac{1}{N-q}\sum_{i=1}^{N-q} v(i)v(i+p)v(i+q) \quad for\ 0 \leq p \leq q \leq N-1$$

$$C'(q,p) = C'(p,q) \quad for\ 0 \leq q < p \leq N-1$$

If $C(p,q)$ has a peak at (p',q') then $\forall i,\ v(i)v(i+p')v(i+q')=1 \Rightarrow C'(p',q')=1$, i.e. $C'(p,q)$ has the same peaks as $C(p,q)$ for $p,q \leq N-1$. For other (p,q) pairs, $v(i+p)v(i+q)=v(i+r)$ for some other shift $r(\neq 0)$, so

$$C'(p,q) = \frac{1}{N-q}\sum_{i=1}^{N-q} v(i)v(i+r) = C'(r)$$

The partial correlation coefficient $C'(r)$ based on an average of N-q products for an *L*-length m-sequence has a mean value of $-1/L$ and a variance of approximately $1/(N-q)$ for $N-q \ll L$, i.e. the variance is inversely proportional to the sequence length used. Thus the principal peaks are identical within the partial window and non peak values have identical expectations with variances dependent on the absolute sample length and *not* on the proportion of the sequence covered by the sample. An example is shown in Figure 4 where the *partial tcf* of a L = 1023 m-sequence with 256 chips intercepted actually gives the same principal peaks as the corresponding *part* of the full *tcf* plot.

The noise theory developed shows that the *tcf* of an m-sequence plus Gaussian noise has the same expected value as the m-sequence *tcf* except for a small bias on its diagonal. Variances depend on the noise variance, with diagonal values three times off-diagonal. The *tcf* of two or more m-sequences in Gaussian noise is more confused by extra spurious peaks due to additional bias terms and increased variances.

Data modulation produces further confusion. Code lengths apart, even short signal intercepts would generally contain at least one polar transition at practical data rates. Even if just one transition occurs half-way through the sequence (at the worst position), much of the *tcf* peak information is preserved.

5. TCF SEARCH FOR SEQUENCE DETECTION

Blind detection of m-sequences may be carried out when fragments of unknown lengths are intercepted. The search relies on the fact that if (p',q') is a peak for an m-sequence length L, then so are $(2^n p' \bmod L,\ 2^n q' \bmod L)$. Possible L values are examined by summing sets of *tcf* values at all such locations within a partial window and comparing against a threshold. Successful (p',q') and L values lead to a set of equations determining the sequences LFSR polynomial in most cases. The process would begin by visiting sequential *tcf* locations (p,q) = (1, 2), (1, 3), ... (2, 3), (2, 4), ..., and the following sum of non-repeated *tcf* values calculated:

$$S = C'(p,q) + C'(2p \bmod L, 2q \bmod L) + ... + C'(2^r p \bmod L, 2^r q \bmod L)$$

where $p = 2^{r+1} p \bmod L$ and $q = 2^{r+1} q \bmod L$. Each doubled location, $(2^i\ p \bmod L,\ 2^i\ q \bmod L)$ for $1 \le i \le r$, is excluded from the future search as its *coset* is represented by the initial location (p,q). Thus each coset present in the partial *tcf* is represented by a single peak, called the 'coset leader'. Clearly, all the peaks of a coset will not generally be present in the partial *tcf*. If this process of modulo doubling gives locations outside the partial tcf $(p',q' \ge N)$ no contribution is made to the sum but doubling is continued, and further values for locations within the partial window summed, until the original (p,q) position is revisited.

As an example we can consider the tcf of a 31-chip sequence [taps 45]. Assuming only a 15x15 partial window is available from intercept, the peak at (2, 5) would be the first one found in the search. The only other coset peak added to S would be at (4, 10); doubling would generate the coset postions (8, 20), (16, 9) and (1, 18) but these lie outside the partial window. Then the peak at (4, 10) would be excluded from the search process and its contribution included at (2, 5), the coset leader in this case.

In addition to reducing dimensionality, the use of coset summation of available *tcf* values will improve the detectability of actual peaks; averaging *tcf* values will reduce the variance. If the search begins with a non-peak location, other positions generated from the modulo doubling process will also be non-peak. Suitable threshold values must be used to discriminate cosets of actual peaks from non-peak groups. Importantly there is generally a dominance of *tcf* values near the diagonal ($p \cong q$) for modulated sequences and the variance of peaks tends to be high for high values of p and q off-diagonal. The results of the process described are relatively insensitive to modulation effects.

6. DISCRIMINATION/DETECTION OF M-SEQUENCES IN NOISE

As, in general, m-sequences have no common *tcf* peaks, detection is easy for high SNRs. In the absence of noise, knowledge of a single peak is sufficient, even for intercepts including only partial sequences. In noise, the detection of a particular sequence is improved by summing *tcf* values at known locations for that sequence. As SNR decreases, more locations are needed to maintain detection rate. Furthermore, the presence of other sequences has little affect on *tcf* values at the original sequence's locations, so the presence of individual sequences may be tested independently.

The power of multivariate discrimination [8] is demonstrated when partial sequences e and f are characterised by the 6-vector:

$$\left[C_e'(p_1,q_1),C_e'(p_2,q_2),C_e'(p_3,q_3),C_f'(p_4,q_4),C_f'(p_5,q_5),C_f'(p_6,q_6)\right]$$

ie *tcf* values from 3 known peak locations for each of the sequences. Five random samples (SNR 0dB) of e+noise and five of f+noise are plotted against the discriminant directions [1, 1, 1, 0, 0, 0] and [0, 0, 0, 1, 1, 1] in Figure 5, showing excellent separation: the horizontal direction represents the *tcf* sum for the 3 e locations and the vertical direction the sum for the f locations. Even when the SNR is reduced to –6dB, separation remains good, with only one sample in ten misclassified.

The above is only a crude illustration. For partial sequences, discrimination is improved if *tcf* values are inversely weighted by their variances, which decrease as p or q approaches the sample length N. Discrimination is more difficult in the presence of data modulation. The following describes how multivariate analysis of variance (MANOVA) may be used to derive optimal discriminant directions (linear combinations of *tcf* values) in the presence of noise and unknown data modulation.

7. DISCRIMINATING MODULATED SEQUENCES WITH 'MANOVA'

Data modulation reduces the number of well defined *tcf* peaks and therefore the separability of sequences. Also, different modulation positions would require different naively estimated discriminant directions. However, the best overall discriminant in the presence of an unknown modulation position may be estimated using the more sophisticated technique of MANOVA to isolate variance/covariance contributions from sequence and modulation position effects and their interaction, leading to best overall discriminant directions in the presence of unknown modulation position. Figure 6 illustrates the separation at 0 dB of two 256-length partial sequences u and v, from 1023-length sequences, modulated after 128 chips, the worst position for discrimination. The two MANOVA derived discriminant directions are linear combinations of *tcf* values dominated by those close to the diagonal (p=q). A figure of 78% successful discrimination was achieved for simulated partial sequences modulated (sign reversed) after 128 values of the 256 sample length, and even higher discrimination performance for different modulation positions.

8. CONCLUSIONS AND FUTURE WORK

The detection of DS/SS communication signals using conventional second-order based processing is currently limited and code sequence characterisation is not possible in practical radio channels. Higher-order statistical processing outperforms power spectral and correlation analysis. Triple correlation however can detect and in limited cases also identify multiple DS/SS transmissions. Discrimination of known m-sequences is possible for short partial windows at low SNRs eg. –6 dB. The effect of data modulation has been investigated and modulation independent discriminants estimated with MANOVA, which has also shown that the same discriminants may be used irrespective of the presence of other m-sequences. The problem of blind detection, where neither the possible sequences nor even sequence lengths are known in advance, has been attempted with the new technique of coset summing for smartly searching the *tcf* space. Generator polynomial estimation may be attempted in most cases from knowledge of a single significant coset sum of *tcf* values. Coset summing is also a powerful technique for reducing *tcf* dimensionality for multivariate discrimination.

It is intended to consolidate the research by extending HOM detection/discrimination to DS chip-codes other than m-sequences and also to explore the robustness of the various *tcf* techniques to the effects encountered in real radio channels including unwanted interference and jamming and also missing data in the intercepted window.

9. REFERENCES

[1] J P Stephens, 'Advances in signal processing technology for electronic warfare' Journal of Electronic Defense, Sep 1995, pp41-50.

[2] P C J Hill and E R Adams, "Emerging Digital Techniques for Detecting and Decoding Spread Spectrum Communication Signals", Proc. of Int. Symposium on Communication Theory & Applications, ISCTA '97, Lake District, UK, 13-18 July 1997.

[3] D L Nicholson, 'Spread spectrum signal design - LPE & AJ systems', Computer Science Press Inc., 1988.

[4] A M S Haisan, 'DS/SS chip coding techniques for enhancing LPI performance', PhD Thesis, Cranfield University RMCS, 1995.

[5] P C J Hill, V E Comley and E R Adams, "Techniques for Detecting and Characterising Covert Communication Signals", Proc. 2nd European Conference on Security and Detection, ECOS '97 London, 28-30 April 1997.

[6] C L Nikias & A P Petropulu 'Higher-order Spectra Analysis - a nonlinear signal processing framework', PTR Prentice-Hall Inc., 1993.

[7] E R Adams, M Gouda and P C J Hill, "Detection & Characterisation DS/SS Signals Using Higher Order Correlation", Proc. IEEE No 4 Int. Symposium on Spread Spectrum Theory & Applications, ISSSTA '96, Mainz, Germany, 22-25 September 1996.

[8] A M Kshirsagar, 'Multivariate Analysis', Marcel Dekker,1978.

[9] E R Adams and P C J Hill, "Detection of DS/SS Signals Using Higher-Order Statistical Processing", Proc. Int. Conference on Acoustics Speech and Signal Processing, ICASSP '97, Munich, Germany, 20-24 April 1997.

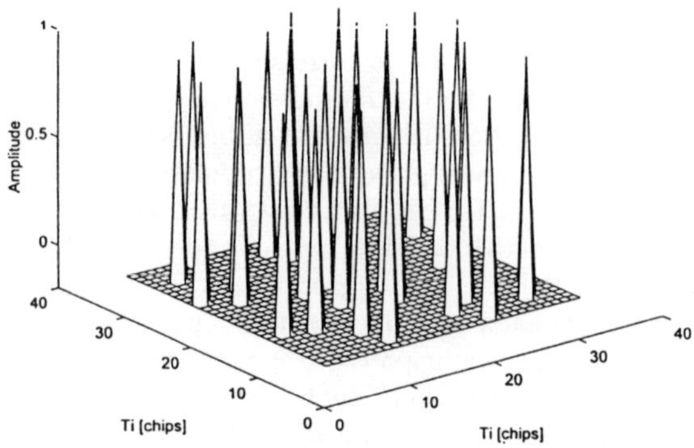

tcf of m-sequence with $g(X)=X^5+X^4+X^3+X^2+1$ **[75]**

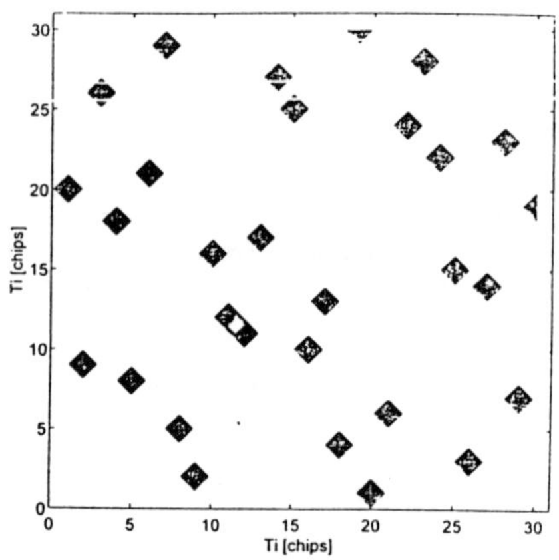

Contour plot of m-sequence [75]

Figure 1

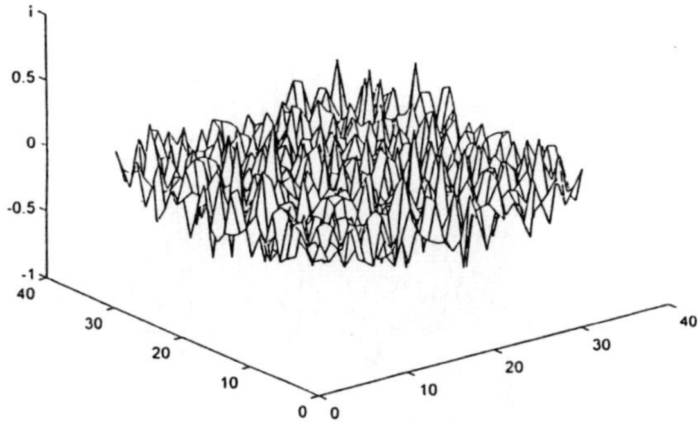

tcf of Gaussian noise

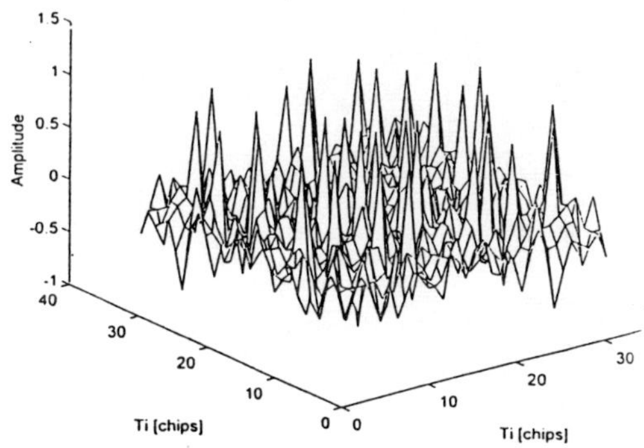

tcf of m-sequence [75]+ noise, SNR=6dB

Figure 2

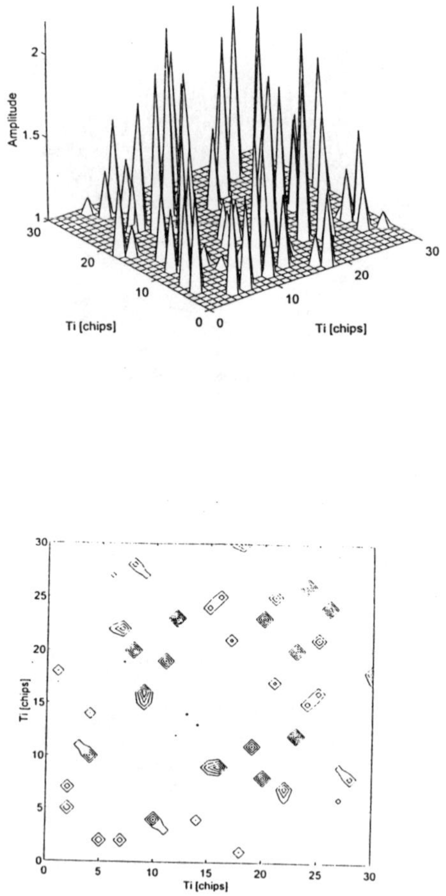

Figure 3 **tcf of superimposed m-sequences [45]+[67] + noise, SNR=6dB**

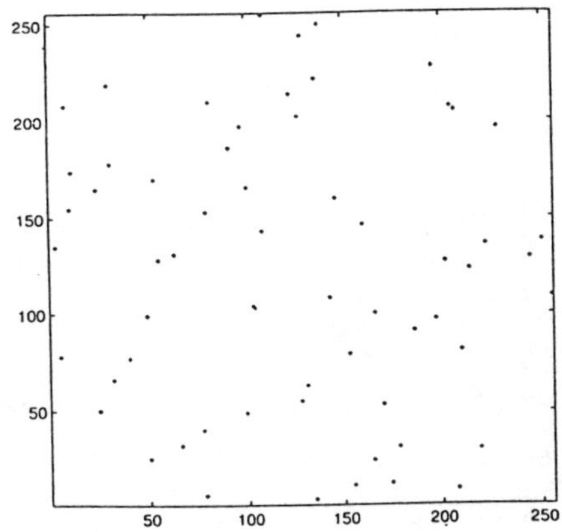

Part of tcf of m-sequence *a*: $g(X)=X^{10}+X^{8}+X^{3}+X^{2}+1$

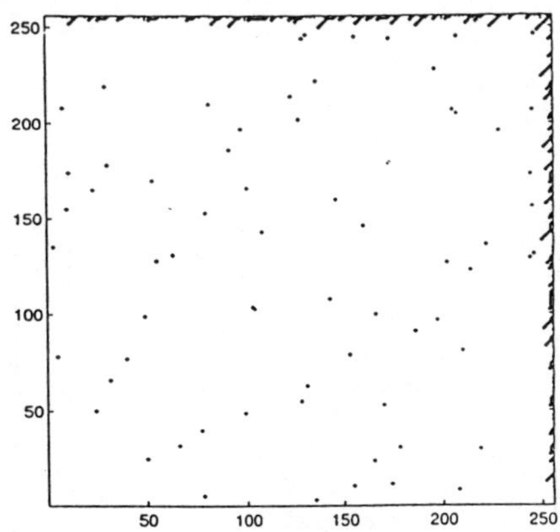

***partial* tcf of *a*, 256/1023**

Figure 4

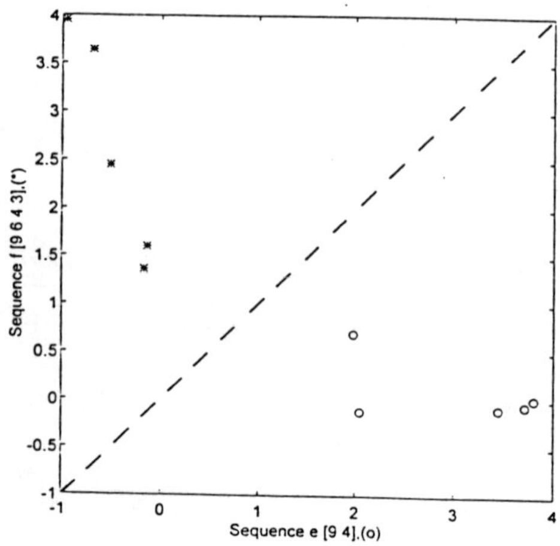

Figure 5

Discrimination between partial sequences *e* & *f*, SNR 0dB

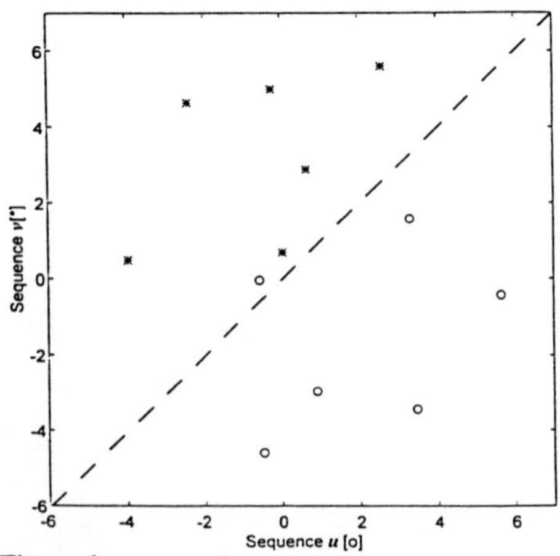

Figure 6

MANOVA based discrimination between partial sequences *u* & *v* data modulated half-way along sample (128/256/1023), SNR 0dB

16. A Reed Solomon Coded Approach to Robust Bandwidth-Efficient Communications

P. Sweeney and E. Husni
Centre for Communication Systems Research
University of Surrey
Guildford GU2 5XH, United Kingdom
Phone: +44 (0) 1483 259123 FAX: +44 (0) 1483 259504
email: p.sweeney@ee.surrey.ac.uk

Abstract

In this paper, we study approaches to the coding of multilevel modulations using only Reed Solomon codes. The selected scheme uses set partitioning of multilevel PSK modulation in conjunction with a rotated, expanded demodulator. By matching configuration of component codes with the channel characteristics, it is found that effective schemes exist both for AWGN and Gaussian channels.

This approach is compared with non-partitioned, Reed Solomon coded schemes and schemes using Reed Muller codes. Simulation results for these schemes show that the chosen scheme compares favourably with the others across a wide range of conditions.

1 Introduction

Much work has been done on design of efficient coded modulation schemes for improving the performance of digital transmission systems since the publication of Ungerboeck's paper [1] for trellis coded modulation (TCM) and Imai and Hirakawa paper [2] for Block Coded Modulation (BCM). Recently, increasing interest for digital mobile radio or indoor wireless systems has led to the consideration of coded modulation design for combating fading channels.

In a number of previous papers [7]-[9], codes were designed for the Rayleigh fading channel so as to maximize their diversity by not using coded modulation techniques for the Gaussian channel.

Here, an alternative approach for combating the Rayleigh fading channel is proposed. The coded modulation system is based on partitioned BCM using Reed Solomon codes which is optimum BCM for the Gaussian channel. By using different configuration of component codes which is matched with the channel characteristic, it is shown that this approach yields a better coding gain over a Gaussian and Rayleigh channel as compared to the previous approach.

There are several reasons for using Reed Solomon codes, such as:

- These codes are maximum distance separable codes, and hence, they make highly efficient use of the redundancy
- Reed Solomon codes are burst error correcting codes, which are suitable for non-Gaussian channels.

- Reed Solomon codes provide a wide range of code rates that can be chosen such that the coded scheme has bandwidth efficiency compatible with the reference uncoded system.

For each code, we use (n, k) Reed Solomon codes over $GF(2^{\nu})$ (where ν is an integer) of length $n = 2^{\nu} - 1$, minimum Hamming distance $(n - k + 1)$ and error correcting capability $(n - k)/2$.

In partitioned BCM, each of the m bits defining an MPSK symbol, where $M = 2^m$, is encoded and decoded by different codecs. The set partitioning principle is applied to define subsets with distances Δ_i, (i = 1 to m) that are nondecreasing with i. Each of the m bits defines a subset and is decoded in multistage decoding schemes. This method was first proposed by Cusack [3], who used Reed Muller codes and a QAM signal set. Later Sayegh [4] generalized Cusack's work to other signal constellations by using known binary codes. Here, we use Reed Solomon codes combined with an MPSK signal set which we call partitioned Reed Solomon Coded Modulation (RSCM).

In section 2, we define a baseline coded modulation approach to which the new coded modulation system is compared. The baseline system uses a Reed Solomon code which is combined with an MPSK signal set using Gray code mapping, this approach is called non-partitioned RSCM.

Specific designs of partitioned RSCM are given in section 3. Performance analysis over the Gaussian channel and the Rayleigh fading channel are dealt with in section 4. Finally, conclusions are given in section 5.

2 Non-Partitioned Reed Solomon Coded Modulation

Here, we address the issue of designing these schemes based on maximizing the time diversity (the effective length) of the code. We consider two methods generalizing this approach.

2.1 Method 1

The first method is that a the symbols of a Reed Solomon code are mapped to the signal points of a 2^m-PSK signal set such that each symbol of the code consists of the concatenation of y channel symbols. In this combination the code rate is chosen such that the rate of the coded scheme is the same as a reference uncoded scheme, usually 2^{m-1}-PSK. This method is based on Jamali and Le-Ngoc's work in [7]. The proposition in [7] indicates that the effective order of time diversity in such a mapping is at least d, the minimum Hamming distance of the Reed Solomon code.

As an example we consider a Reed Solomon (63, 42) code, defined over $GF(2^6)$, combined with 8-PSK, and hence, each code symbol consists of two concatenated 8-PSK symbols. The rate of the code is 2/3 which translates into 2 bits/s/Hz throughput, equivalent to that of uncoded QPSK.

2.2 Method 2

The second possibility is that a combination of MPSK signal and Reed Solomon code of appropriate rate may be chosen with no simple relationship between symbol and constellation sizes. For example, a (30, 15) Reed Solomon code, defined over $GF(2^5)$, could be combined with QPSK signaling. The overall coded QPSK throughput is the same as that of uncoded BPSK, i.e., 1 bit/s/Hz.

3 Partitioned Reed Solomon Coded Modulation

The BCM encoder consists of m Reed Solomon codes, called RS component codes; this is illustrated in Fig. 1. Binary digits are assigned to each point in the signal space according to Ungerboeck's set partitioning scheme [1]. We now form an array of v·n columns and m rows, where v denotes bit length of one symbol and n coded symbol length.

$$\begin{matrix} b_1^{(1)} & b_2^{(1)} & \cdots & b_{v\cdot n}^{(1)} \\ b_1^{(2)} & b_2^{(2)} & \cdots & b_{v\cdot n}^{(2)} \\ \cdots & \cdots & \cdots & \cdots \\ b_1^{(m)} & b_2^{(m)} & \cdots & b_{v\cdot n}^{(m)} \end{matrix}$$

where $b_j^{(i)} \in \{0,1\}$ for $1 \le j \le v\cdot n$. Each column of the array will correspond to one point in the signal space, with the bit in the first row corresponding to the leftmost digit and the bit in the last row corresponding to the rightmost digit in the representation of the signal space points. The array will be transmitted one column at a time, each column being represented by the corresponding signal space point.

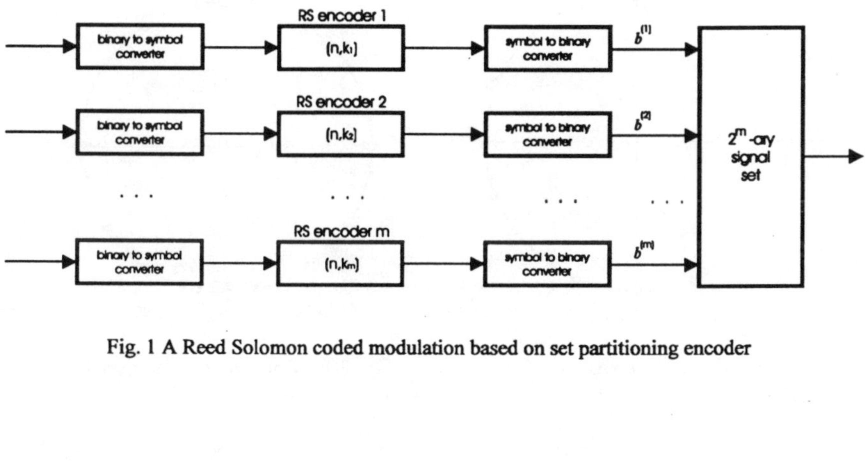

Fig. 1 A Reed Solomon coded modulation based on set partitioning encoder

The array contains m·n·v bits of which v·n·m·R_c bits are information bits. Denoting the number of information bits in ith row by v·k_i, we can write

$$v \cdot k_1 + v \cdot k_2 + \cdots + v \cdot k_m = v \cdot m \cdot n \cdot R_c$$

For a given rate R_c, the values of k_i's are chosen subject to the above conditions in such a way as to maximize the minimum Euclidean distance between the codewords of the code.

3.1 Multistage Decoding

A multistage decoding approach has been used for partitioned RSCM. Multistage decoding of multilevel trellis modulation codes has been recently studied and analyzed in a number of papers [5, 6]. The main case of interest here is using a block encoder and block decoding algorithm for each component code of a multilevel modulation code.

In principle a coded modulation scheme should be decoded using minimum Euclidean distance between the received sequence and the overall code. It is therefore desirable to employ a finer quantization on the received levels from the demodulator than is employed in the modulator, in other words we need soft decision demodulation. Soft decision decoding methods for Reed Solomon codes are not, however, practicable for most of the codes considered here. It was therefore decided to use a multistage decoding scheme and attempt to utilize the error detection capability of the component RS codes. Here, however, we encounter another difficulty in that the decisions on any bit depend on knowledge of the other bits in that transmitted symbol. Moreover, the points of an expanded MPSK constellation lie on the decision boundary for one of the bits. For example, in a set-partitioned 8-PSK constellation the points lying between the QPSK points each lie on the decision boundary for one of the QPSK bits. Thus for any such received points, a decoding failure with detected errors would leave other bit values unresolved.

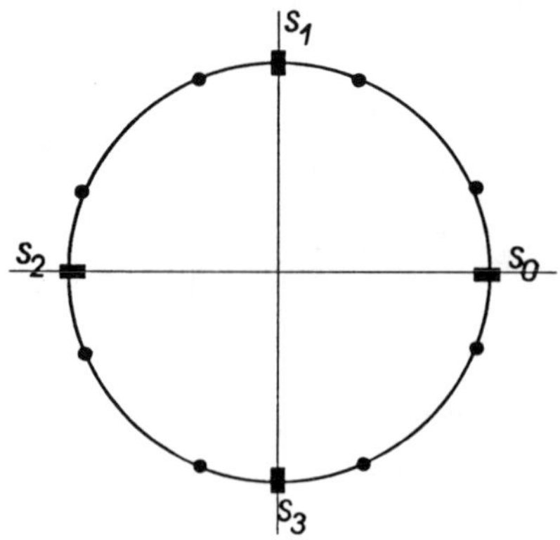

Fig. 2(a). Rotated 8-PSK signal set for a QPSK modulator

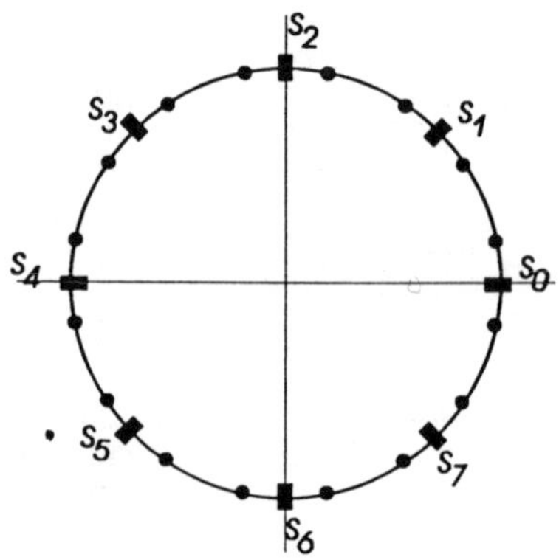

Fig. 2(b). Rotated 16-PSK signal set for an 8-PSK modulator

To solve this problem, we used in the receiver, a rotated 2^{m+1}-PSK detector with a 2^m-PSK modulator. This is illustrated in Figure 2(a), a detector of a QPSK modulator, and Figure 2(b), a detector of an 8-PSK modulator. This ensures that the received level does not fall on a decision boundary when decoding any of the bits in the symbol. It can be seen that there are two signal points of the rotated 2^{m+1}-PSK signal set in each signal point's region of the 2^m-PSK signal set.

The block diagram for multistage decoding of partitioned RSCM using 8-PSK is illustrated in Fig. 3. Let L be the labeling for a two dimensional signal set S with 2^m signal points. Let $C \equiv C^{(1)}C^{(2)}\cdots C^{(m)}$ be an m-level code of length n over L with m component codes where $C^{(i)}$ is a code over the sublabeling $L^{(i)}$. In multistage decoding of C, component codes are decoded one at a time, stage by stage. The decoded information at each stage is passed to the next stage.

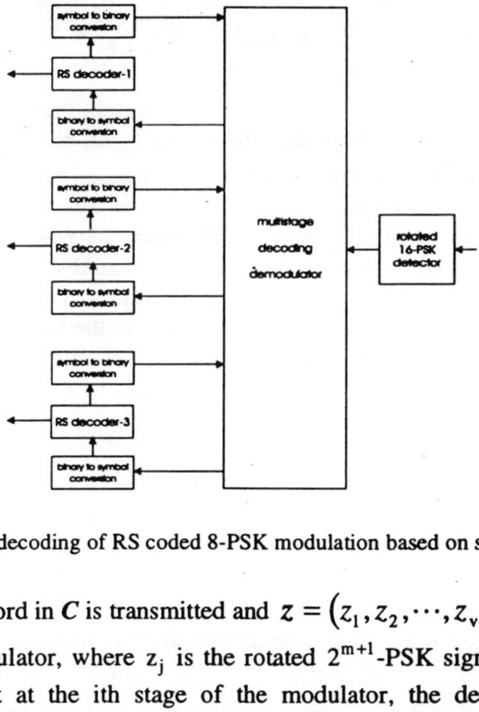

Fig. 3. Multistage decoding of RS coded 8-PSK modulation based on set partitioning.

Suppose a codeword in C is transmitted and $z = (z_1, z_2, \cdots, z_{v \cdot n})$ is the received sequence at the modulator, where z_j is the rotated 2^{m+1}-PSK signal point. Given a hard decision output at the ith stage of the modulator, the decoder performs a decoding for $C^{(i)}$. If the decoding is successful, the decoder puts out a decoded codeword $\mathbf{v}_{\mathbf{D}}^{(i)}$ which is in $C^{(i)}$. Otherwise, $\mathbf{v}_{\mathbf{D}}^{(i)}$ is a null array.

The demodulator makes a hard decision as follows. For the given received point z_j and decoded sublabels $\mathbf{v}_{\mathbf{D}}^{(g)} \in \mathbf{L}^{(g)}$ with $1 \le g < i$ and $i < g \le m$, find the label $v = \mathbf{v}_{\mathbf{Dj}}^{(1)} \cdots \mathbf{v}_{\mathbf{Dj}}^{(i-1)} \mathbf{v}^{(i)} \mathbf{v}_{\mathbf{Dj}}^{(i+1)} \cdots \mathbf{v}_{\mathbf{Dj}}^{(m)}$ in L with $\mathbf{v}_{\mathbf{Dj}}^{(1)} \cdots \mathbf{v}_{\mathbf{Dj}}^{(i-1)}$ as a prefix and

$\mathbf{v}_{\mathbf{Dj}}^{(i+1)}\cdots\mathbf{v}_{\mathbf{Dj}}^{(m)}$ as a suffix such that the norm $\left\|z_j - s(v)\right\|^2$ is minimized, where $s(v)$ denote the signal point in 2^m-PSK represented by $\boldsymbol{v}$. If there is no decoded sublabels $\boldsymbol{v}$ is a null string.

Either one or two iterations may be used for decoding. In the first iteration, level 1 is first decoded, then level 2 is estimated based on the result of level 1. Finally level 3 is determined based on levels 1 and 2. For the second (optional) iteration, level 1 is again decoded using the result of levels 2 and 3 from the first iteration. The corresponding is done for levels 2 and 3.

3.2 Distance Considerations

The minimum squared Euclidean distance of binary block coded 8-PSK scheme is obtained as

$$D_{\min}^2 = \min\left(\Delta_0{}^2 d_1, \Delta_1{}^2 d_2, \Delta_2{}^2 d_3\right) \tag{1}$$

Thus in binary block coded scheme design, the minimum Hamming distance of each component code can be determined by setting $\Delta_0{}^2 d_1 \cong \Delta_1{}^2 d_2 \cong \Delta_2{}^2 d_3$.

The optimum values for minimum Hamming distance of each component code of partitioned RSCM can not be determined. Firstly Reed Solomon codes are non-binary codes so minimum Hamming distance is not a binary measure and Equation (1) is no longer valid for this code. Secondly in multistage decoding the first stage decoding gives an output codeword to the second stage decoding, and so on. Thus the second stage decoding depends on and takes advantages from the first stage decoding, and so on. Thirdly Equation 1 is valid for a Gaussian channel only. Consequently, equation 1 can not be used to determine exactly the minimum Hamming distance of each component code.

4 Performance

In this section, the performance of partitioned RSCM over the Gaussian channel and the Rayleigh fading channel are compared with non-partitioned RSCM and RMCM using computer simulations.

BCM using Reed Muller codes require the same approaches. At the first approach, Reed Muller code having code rate $R_c = {}^{m}\!/_{(m+1)}$ combined with a 2^{m+1}-PSK signal set. In this combination the code rate is chosen such that the rate of the coded scheme is the same as the uncoded one (2^m-PSK). At the second approach, Reed Muller codes are used for component codes. It uses similar multistage decoding procedure for decoding the received codewords.

4.1 Gaussian Channel

Reed Solomon codes provide a wide range of code rates, thus there are many configurations of component codes for RSCM based on set partitioning. Table 1 gives a list of good codes for each specified coded symbol length for coded 8-PSK modulation and Table 2 for coded QPSK modulation.

level	k	
	n = 63	n = 127
1	7	11
2	59	119
3	61	125
throughput (bits/s/Hz)	2.01	2.007
information length (bits)	(7+59+ 61)·6 = 762	(11+119 +125)·7 = 1785

Table 1. Configuration of good codes for RS coded 8-PSK modulation for coded symbol length of 63 and 127.

level	k	
	n = 31	n = 63
1	7	11
2	25	53
throughput (bits/s/Hz)	1.03	1.01
information length (bits)	(7+25)·5 = 160	(11+53)·6 = 384

Table 2. Configuration of good codes for RS coded QPSK modulation for coded symbol length of 31 and 63.

As previously mentioned for the binary block coded scheme, the minimum Hamming distance of each component code can be determined using eq. 1. This however does not hold for multilevel codes; for example, for the best multilevel code of length 63 from Table 1

$$\Delta_0^2 \cdot d_1 = 0.586 \cdot 57 = 33.4$$
$$\Delta_1^2 \cdot d_2 = 2 \cdot 5 = 10$$
$$\Delta_2^2 \cdot d_3 = 4 \cdot 3 = 12$$

From the results we can see that eq. 1 is not valid for the best multilevel code of each code rate and code symbol length.

Figs. 4-5 show block error probability for Reed Solomon coded QPSK modulation and coded 8-PSK modulation. In these figures, the error performances are compared with those of some uncoded reference modulation systems for transmitting the same (or almost the same) number of information bits.

We can see that for coded QPSK modulation, partitioned RSCM has improvement on non-partitioned RSCM by an amount approximates equivalently to doubling code length. For coded 8-PSK modulation at block error probability of 10^{-4}, set partitioning produces 1 dB more coding gain than doubling the code length.

Fig. 6 shows bit error rates of various coded modulation schemes for coded QPSK modulation. Reed Muller codes used here are however more complex to decode than the Reed Solomon codes used. At bit error probability of 10^{-4}, partitioned RSCM has 0.75 dB more coding gain than non-partitioned RSCM with the same code length. It appears that for coded QPSK modulation, Reed Muller coded modulation based on set partitioning is worse than Reed Muller coded modulation not based on set partitioning. For example, at bit error rate 10^{-4}, Reed Muller coded modulation based on set partitioning, length 256, is 0.5 dB worse than Reed Muller coded modulation not based on set partitioning, length 128.

Fig. 7 shows bit error rates of various coded modulation schemes of coded 8-PSK modulation. For partitioned coded modulation, it turns out that all codes have about the same performances. At a bit error rate of 10^{-4}-10^{-5}, partitioned coded modulation has at least 1 dB coding gain over non-partitioned coded modulation for the same code length.

In this channel, it appears that one iteration and two iterations multistage decoding have about the same performances. In [6] it was claimed that in the AWGN channel, two iterations will give better performance than one iteration if there is an interleaver between the coded bits of all stages or we use modified generalized-minimum-distance-decoding. They used a three dimensional block interleaver for three levels coded modulation with memory cells as many as coded bit length in every dimension. Every memory cell contains one coded bit from every level. The coded bits of the first level are always written in the cells corresponding to the direction of the first dimension. The coded bits of the second level are written in according to the direction of the second dimension. The corresponding is done for the third level.

They used binary convolutional codes for the first and second levels and a single parity check for the third level. If we use the Viterbi algorithm for decoding, the estimated information bit sequence is liable to contain error bursts. This sequence is re-encoded and fed into the decoder of the next level. Thus, the re-encoded sequence also contains error bursts. If this decoder also uses the Viterbi algorithm then it is very sensitive to these error bursts, because the algorithm is designed to deal with independent errors in the input stream.

In order to avoid this error propagation effect, they introduced interleaving between the coded bit streams of each level as explained above. The interleaving spreads the re-encoded bit streams of any two decoders for the third decoder. So it will be hardly influenced by error bursts from any of the other two decoders.

Reed Solomon codes have powerful error detection capability which are different from binary convolutional codes. In the simulations if the decoder detected the errors but it could not correct the errors, the decoder passed the input sequence to the decoder output. Therefore there is very low probability of an error propagation effect because of undetected errors.

4.2 Rayleigh Fading Channel

In this subsection, we analyse the error performance of partitioned RSCM compared with non-partitioned RSCM and RMCM over the nonselective slow Rayleigh fading channel. Here 'slow' means that the fading bandwidth is small compared to the signal bandwidth so that the receiver will be able to track the phase variations.

For all schemes of Reed Solomon coded modulation, the code symbols are interleaved before modulation in order to destroy the memory of the fading channel.

Figs. 8 and 9 show bit error rates of various coded modulation schemes for coded QPSK and 8-PSK modulation. We can see that in this channel, RSCM schemes have a large coding gain to Reed Muller coded modulation schemes. We also see that the error performances of partitioned RSCM using 8-PSK are better than non-partitioned RSCM at high BER, and finally they become the same at low BER; and the error performances of partitioned RSCM using QPSK are better than those of non-partitioned RSCM. Therefore, the error performances partitioned RSCM are never worse than those of non-partitioned RSCM for the same code length.

This scheme is different from the Gaussian channel in that good configurations have the same component codes. Therefore, it seems that all levels have the same distance. In other words, the fading phase is a uniformly distributed random process. By matching configuration of component codes to the channel characteristic, it is shown that partitioned RSCM can be robust codes for the Rayleigh fading channel.

The error performances of RSCM not based on set partitioning using the first approach whose code symbol consists of one channel symbol can be seen in [7]. These schemes are not simulated because they have limited configurations.

Figs. 10-11 compare error performances between one iteration and two iterations multistage decoding. We can see that one and two iterations have different good codes. The good codes of one and two iterations differ by 0.5 dB at a bit error rate of 10^{-4}.

In Fig. 10, a good code of one iteration decoding has component codes RS(31,11) and RS(31,21), and in Fig. 11 the good code has component codes RS(63,19), RS(63,51) and RS(63,57).

Therefore we can conclude that in this channel, two iteration multistage decoding must be used. This is because the good codes have the same component codes, thus the first level component code also needs decoded codewords of other levels.

5 Conclusions

In this paper, partitioned Reed Solomon coded modulation (RSCM) is proposed to provide bandwidth-efficient communications both for the additive white Gaussian noise channel and a Rayleigh fading channel. A novel, rotated, expanded MPSK receiver was combined with multistage, hard decision decoding to obtain performance that compares favourably with other schemes. The flexibility in application to diferent types of channel is achieved by matching configuration of component codes with the channel characteristics. It was also found that a second iteration of the decoding was beneficial in Rayleigh conditions, but not for the Gaussian channel.

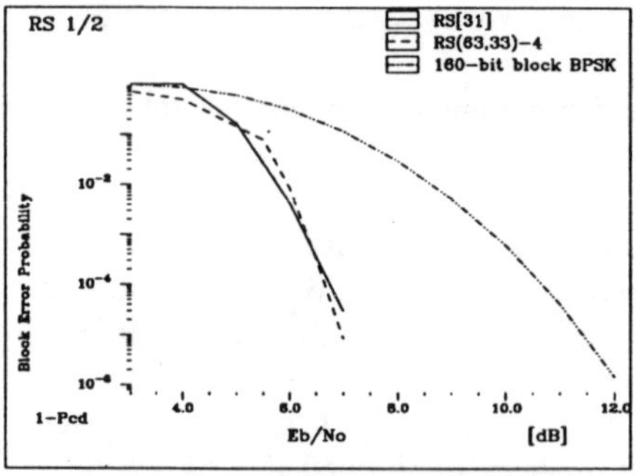

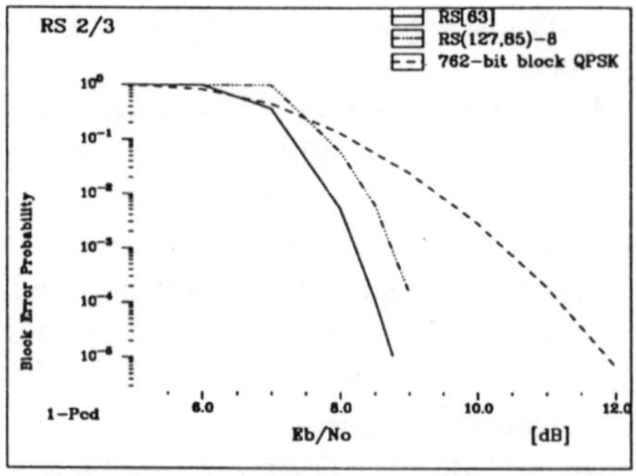

Fig. 4. Error performances of coded QPSK modulation; partitioned RSCM **RS[63]** of length 63 listed in Table 2 and non-partitioned RS(127, 65) with information-
bit length of 65·7 = 455.

Fig. 5. Error performances of coded 8-PSK; partitioned RSCM **RS[127]** of length 127 listed in Table 1 and non-partitioned RS(255,171) with information-bit length of 171·8 = 1368.

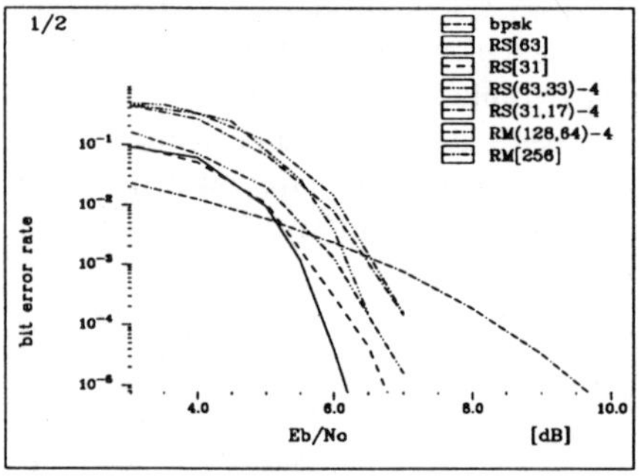

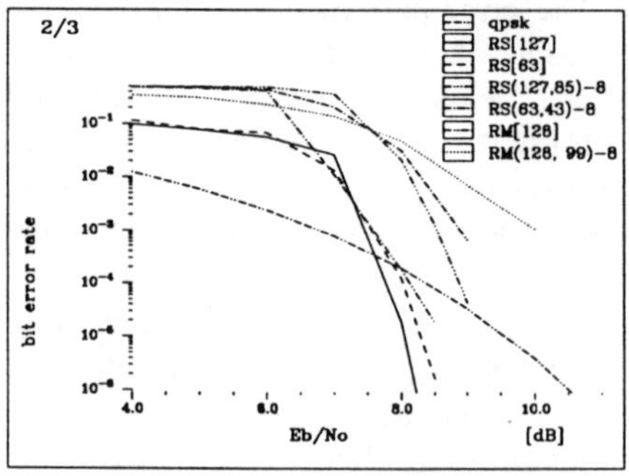

Fig. 6. Error performances of coded QPSK modulation; partitioned RSCM: **RS[63]** of length 63 listed in Table 2, **RS[31]** of length 31 listed in Table 2, non-partitioned: RS(63,33), RS(31,17), and partitioned RMCM **RM[256]** of length 256, non-partitioned 3rd-order RM(128,64).

Fig. 7. Error performances of coded 8-PSK modulation; partitioned RSCM: **RS[127]** of length 127 listed in Table 1, **RS[63]** of length 63 listed in Table 1, non-partitioned: RS(127,85), RS(63,43), and partitioned RMCM **RM[128]** of length 128, non-partitioned 4th-order RM(128,99).

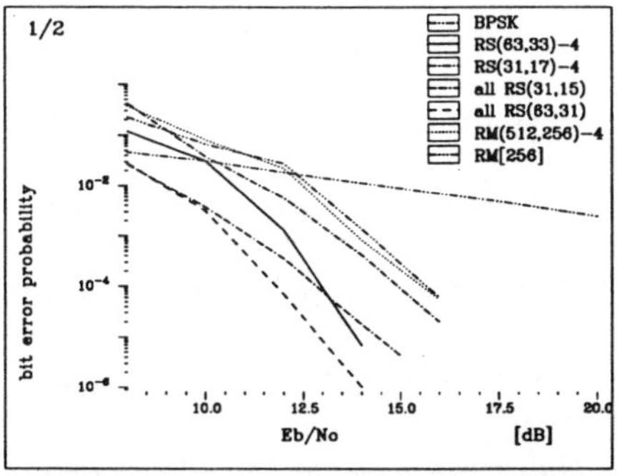

Fig. 8. Error performances of coded QPSK modulation over a Rayleigh fading channel; partitioned RSCM: **all RS(63,31)** of length 63 whose component codes are all RS(63,31), **all RS(31,15)** of length 31 whose component codes are all RS(31,15), non-partitioned: RS(63,33), RS(31,17), and partitioned RMCM **RM[256]** of length 256, non-partitioned 4th-order RM(512, 256).

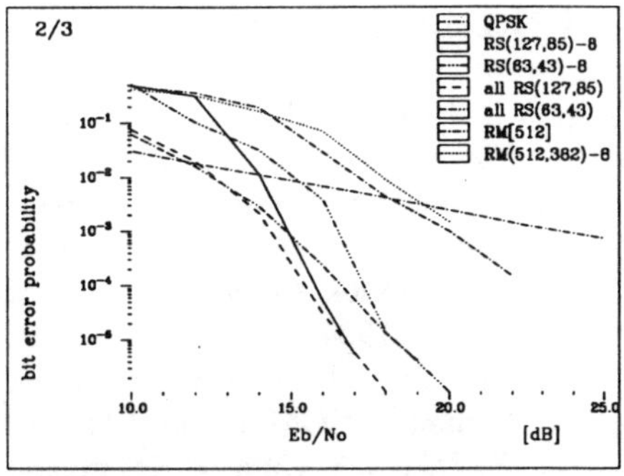

Fig. 9. Error performances of coded 8-PSK modulation over a Rayleigh fading channel; partitioned RSCM: **all RS(127,85)** of length 127 whose component codes are all RS(127,85), **all RS(63,43)** of length 63 whose component codes are all RS(63,43), non-partitioned: RS(127,85), RS(63,43), and partitioned RMCM **RM[512]** of length 512, non-partitioned 5th-order RM(512, 382).

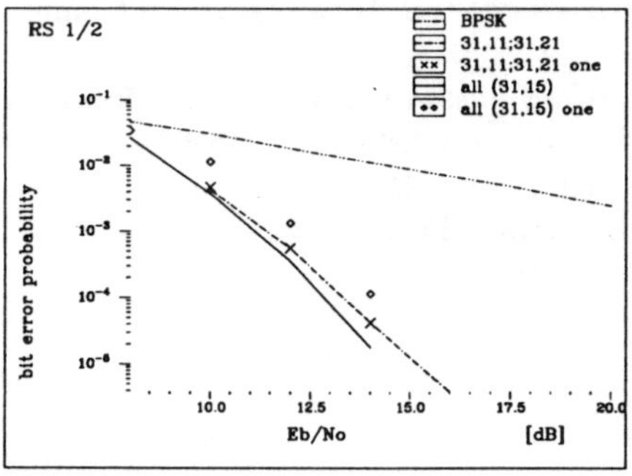

Fig. 10. Error performances over a Rayleigh fading channel of RS coded QPSK modulation based on set partitioning using two iterations multistage decoding: **all (31,15)** of length 31 whose component codes are all RS(31,15), **31,11;31,21** of length 31 whose component codes are RS(31,11) and RS(31,21), and one iteration multistage decoding: **all (31,15) one, 31,11;31,21 one**.

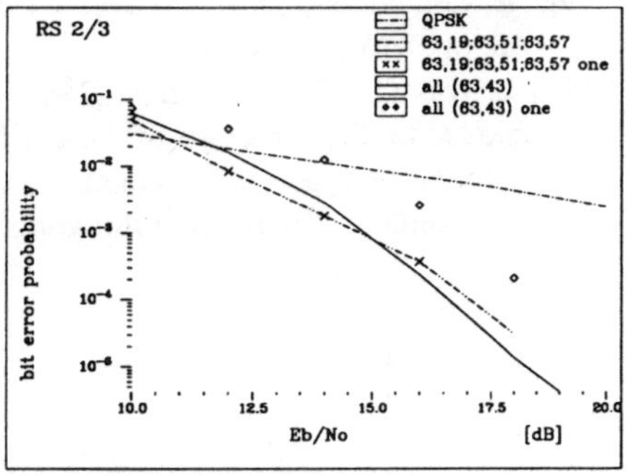

Fig. 11. Error performances over a Rayleigh fading channel of RS coded 8-PSK modulation based on set partitioning using two iterations multistage decoding: **all (63,43)** of length 63 whose component codes are all RS(63,43), **63,19;63,51;63,57** of length 63 whose component codes are RS(63,19), RS(63,51), and RS(63,57), and one iteration multistage decoding: **all (63,43) one, 63,19;63,51;63,57 one**.

References

[1] G. Ungerboeck, *Channel coding with multilevel/phase signals*, IEEE Trans. Inf. Theory, IT-28, pp. 55-67, Jan. 1982.

[2] H. Imai and S. Hirakawa, *A new multilevel coding method using error correcting codes*, IEEE Trans. Inf. Theory, IT-23, pp. 371-376, May 1977.

[3] E.L. Cusack, *Error control codes for QAM signaling*, Electron. Lett., 20, pp. 62-63, Jan. 1984.

[4] S.I. Sayegh, *A class of optimum block codes in signal space*, IEEE Trans. Commun., COM-30, pp. 1043-1045, Oct. 1986.

[5] T. Takata, S. Ujita, T. Kasamiand S. Lin, *Multistage decoding of multilevel block M-PSK modulation codes and its performance analysis*, IEEE Trans. Inf. Theory, 39, pp. 1204-1218, July 1993.

[6] T. Woerz and Fazel, *Comparison of different decoding strategies for block coded modulation*, in Proceedings of 1992 URSI Int. Symp. Signals, Sys. and Electronics, Paris, pp. 117-120, Sept., 1992.

[7] S.H. Jamali, *Coded modulation techniques for fading channels*, Kluwer Academic Publishers, 1994.

[8] E. Biglieri, G. Caire, G. Taricco and J. Ventura-Traveset, *Coding and modulation for the fading channel: the quest for robustness*, in Proceedings of the Fifth ESA Intl. Workshop on DSP techniques Applied to Space Communication, Barcelona, pp. 8.1-89, Oct., 1996.

[9] E. Zehavi, *8-PSK trellis codes for a Rayleigh channel*, IEEE Trans. on Comm., 40, pp. 873-884, May 1992.

[10] E. Husni and P. Sweeney, *Block coded MPSK modulation using Reed Solomon codes*, in Proceedings of the Fifth ESA Intl. Workshop on DSP techniques Applied to Space Communication, Barcelona, pp. 1-8, Oct., 1996.

[11] P. Sweeney, *Error control coding: an introduction*, Prentice Hall, 1991.

CHAPTER 5

Communications Systems and Networks

17. Simulation of Non-Linear Amplifiers with Complex Modulated Signals for use in Space

R. Carter
S. Sonander
B. Honary
Lancaster Communications Research Centre
Lancaster University, UK

Abstract

The Authors investigate the validity of Travelling Wave Tube Amplifier (TWTA) models when applied to satellite communication systems. Techniques of intermodulation analysis of power amplifiers are discussed and compared for different system configurations. Both three and eight carrier schemes are investigated. A technique is described which optimises the modulation parameters governing the downlink signal of a deep space transponder.

1. Introduction

The accurate simulation of non-linear communication systems has long been a problem. In many deep space satellite systems, large output powers are essential and Travelling Wave Tube Amplifiers (TWTA) are commonly used as the final power amplifiers [1]. Depending upon the type of modulation scheme chosen for the communication system, the choice of the TWTA and its operating point is often a critical optimising factor. The requirements of any satellite communication system need to be addressed early, as mistakes can be costly. Certain trade-offs exist. Generally, the requirements are for high efficiency and low weight, minimal heat dissipation, low bit error rate and circuit complexity. The choice of mean power to give a trade-off between the desired performance and the target of high efficiency also needs careful consideration. It is therefore highly desirable to simulate the entire system in order to achieve optimum performance.

In this paper we describe techniques for determining the multi-carrier performance of a TWTA based upon the single-carrier transfer characteristics. Although the performance of a TWTA varies with frequency it can often be described, to a good approximation, by the single-carrier transfer characteristics at the band centre. The prediction of the multi-carrier transfer characteristics from this information is highly desirable because it is then possible to simulate many complex communication systems in a simple manner.

The techniques described in the paper are applied to an example of deep space satellite communication system whose final stage is a non-linear TWTA. The satellite communication system requires the modulation of three separate information signals onto a single carrier wave. Following the non-linear amplification of the signal at 8.4GHz, it will be shown how the system can be optimised in terms of bit error rate, sub-carrier frequencies and modulation parameters.

2. Multi-carrier simulation of a TWTA

A method for the intermodulation analysis of power amplifiers has been described by Gatti et al. [2].

"Imal-2 predicts the multi-carrier performance of an amplifier based on the single-carrier characteristics of the amplifier (AM/AM., AM/PM and efficiency versus Pin). this approach is valid only for quasi-memory-less non-linear systems. Normally a well designed microwave amplifier falls within this category. Basically IMAL utilizes an input signal defined in the frequency domain. An initial random phase value is assigned to each spectral line and using inverse Fast Fourier Transform (FFT) the spectrum is converted in the time domain. As we are interested in the intermodulation falling in the amplifier band, the band-pass input signal is then converted to a complex low-pass equivalent one (complex envelope) and this is distorted by using the non-linear characteristic of the microwave amplifier. A liner interpolation routine is used to distort the signal and afterwards the signal is transformed back to frequency domain (FFT) to obtain the desired characteristics. In general the phase value of the spectral lines in a multi-carrier signal are statistically independent. Hence the output spectrum has to be averaged over a different random phase setting of the input signal and, generally, for standard amplifier non-linear responses the convergence is obtained after 50 to 500 interations."

Detailed analysis of the non-linear behaviour of a travelling-wave tube reveals complex behaviour in which the trajectories of individual electrons may cross over each other so that their positions and velocities are no longer single-valued functions of the r.f. phase at which they entered the tube. Such cross-overs normally begin to occur at aproximately the input signal level for which the transfer characteristics are perceptibly non-linear. It is, therefore, not self-evident that the non-linear performance of a TWTA can be represented adequately by an analysis based on its single-carrier transfer characteristics.

In order to establish the extent to which this method is valid a comparison was made between results obtained using IMAL-2 and those from a Lagrangian analysis using a computer simulation which included multiple sinusoidal signals (LSM)[3]. These could be input signals whose amplitude and phase were specified independently, intermodulation products and harmonics of the input signals. Because the computation time required for this method is long the comparisons were restricted to three cases

- Two equal carriers
- Three equal carriers arranged as shown in Figure 1 with phases varied in 60 degree steps
- Eight equal carriers arranged as shown in Figure 2.

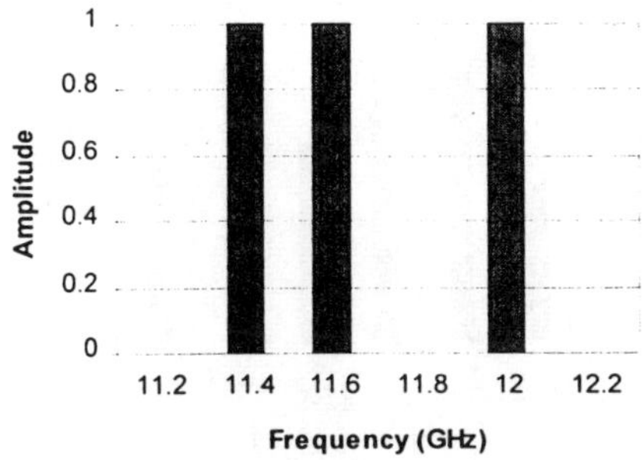

Figure 1 - Three Carrier Scheme

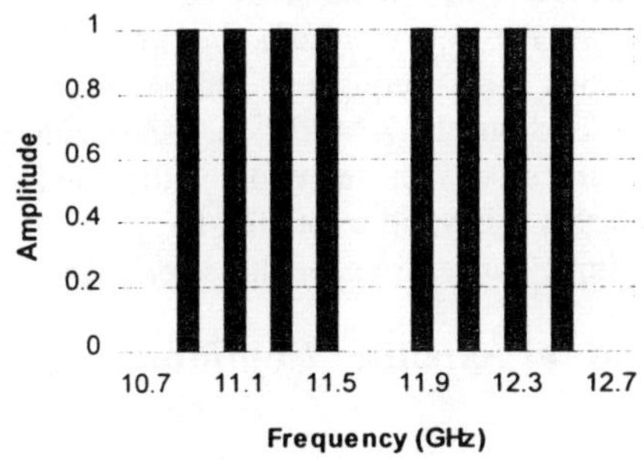

Figure 2 - Eight Carrier Scheme

In the case of two carriers all possible phase relationships exist within one period of the difference frequency. For eight carriers it was only possible to compute the single case for which all the carriers were in phase with each other initially. For both these cases the comparison with IMAL-2 results showed good agreement. The eight-carrier case was an approximation to the method of specifying the performance of an amplifier by its noise power ratio (NPR) [4] in which the input signal is band-limited white noise with a central notch and the intermodulation power appearing in the notch is measured.

The results for the three-carrier simulation are shown in Figure 3 with those obtained from IMAL-2 for comparison. The results show the average of all 36 phase cases at a drive level approximately 3dB back-off from saturation.

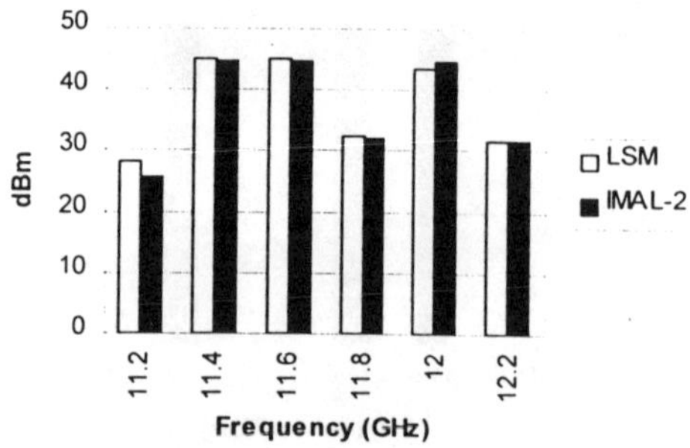

Figure 3 - Three Carrier Results

There is good agreement between the results obtained by the two methods. The so the slight differences between the two sets of results are probably the result of variations in the transfer characteristics with frequency. The AM/AM curves differed little across the frequency band used but the AM/PM characteristics showed a variation in phase lag of around 10 degrees between the ends of the band at this level of output back-off. It was concluded that it is valid to predict the performance of a TWTA with complex signals from its single-carrier transfer characteristics.

3. An Example of a Satellite Communication System with a TWTA

The deep space transponder exciter stage that has been simulated is described in this section [5]. The overall system is described and then the simulation method is explained. Finally, the results are presented.

3.1 The Deep Space Satellite System

The satellite has three information signals, which are to be modulated onto the same carrier wave. These three information signals are:

1) The telemetry information (TM)
2) The ranging signal (Rng)
3) The ΔDOR signal

The TM information signal is assumed to be memoryless random binary data. The Rng signal is a known fixed sequence of binary information and is used to calculate the distance of the satellite from the ground station. The ΔDOR signal is the addition of two sinusoidal waveforms and is used to determine the angular resolution of the satellite. The TM and Rng signals amplitude and phase modulate a square wave and a sine wave subcarrier respectively. The sum of these two subcarriers is further used to phase modulate a sinusoid which in turn is amplitude modulated by the ΔDOR signal. The modulation scheme employed by the system is given as equation 1

$$\Delta DOR(t)\sin\left\{\omega_c + \Phi_{TM} TM(t) square\left(\omega_{TM} t + \varphi_{TM}\right) + \Phi_{Rng}\cos\left(\omega_{Rng} t + Rng(t)\right)\right\} \qquad (1)$$

where $\Delta DOR(t)$, $TM(t)$ and $Rng(t)$ are the ΔDOR, TM and Rng signals respectively. Φ_{TM} and Φ_{Rng} are the modulation indices, ω_{TM} and ω_{Rng} are the TM and Rng subcarrier angular frequencies. Finally, ω_c is the carrier angular frequency which is in the X-band at 8.4GHz. Various levels of filtering and frequency upconversions are involved in the modulation process. The final stage of the exciter stage is the amplification by the TWTA.

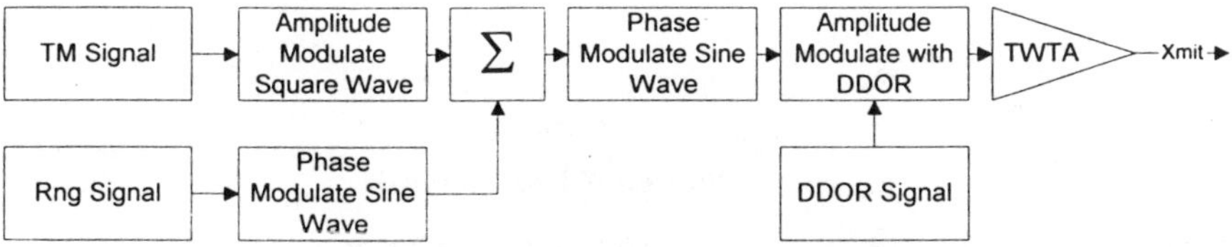

Figure 4 - Exciter Stage Schematic

3.2 Simulation Method

The simulation of the exciter stage has been split into three sections. The first section covers the simulation of the modulation scheme used in the exciter stage. The second section covers the simulation of the TWTA and the third section describes the technique used to optimise the bit error rate of the system.

3.2.1 *Simulation of Modulation Scheme*

The first step in the simulation of the modulator involves the generation of the TM and Rng signals. The signals are represented in the time domain with a sampling frequency approximately 4 times the highest frequency. The TM signal amplitude modulates a square wave and the Rng signal phase modulates a sine wave. The two signals are then added together and filtered using a 4th order Butterworth filter [6]. The resulting signal phase modulates a sine wave. The frequency upconversions are achieved by increasing the sample frequency using Fourier transform techniques [7]. At this point the ΔDOR signal is generated and used to amplitude modulate the signal. One final frequency upconversion performed, increasing the carrier wave frequency to the required 8.4GHz.

3.2.2 *The Simulation of the TWTA*

For the purpose of simulation, the TWTA specification was assumed to have a constant frequency response. The transfer characteristic of the TWTA is shown as output power as a function of input power as Figure 5

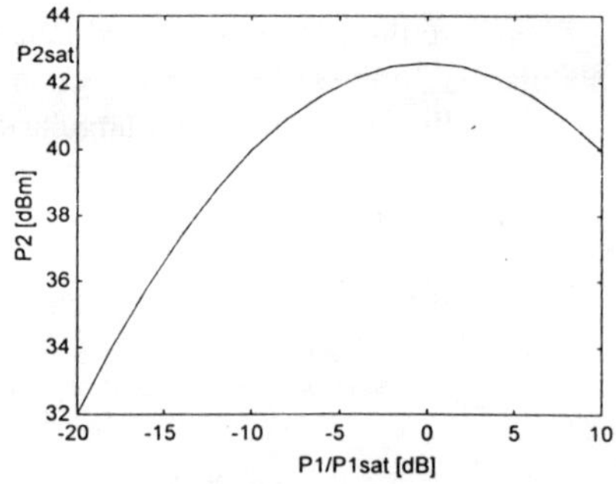

Figure 5 - Pin/Pout Characteristic

and the phase response as a function of input power as Figure 6

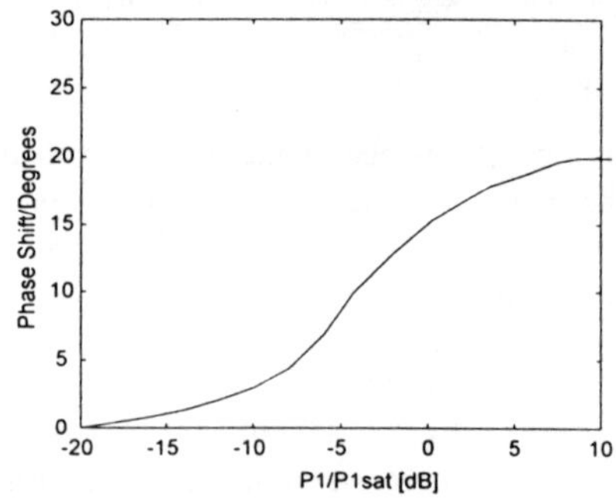

Figure 6 - Pin/Phase Characteristic

The relationship between input voltage and output voltage can be derived from the power relationship and is given as

$$V_{out} = F_{V_{out}}\left\{V_{in}\right\} \tag{2}$$

and the relationship between input power and phase is given as

$$\varphi_{shift} = F_{\varphi_{shift}}\left\{V_{in}\right\} \tag{3}$$

Thus, the actual voltage output from the TWTA is given as

$$V = F_{V_{out}}\left\{V_{in}\right\}.\exp\left(jF_{\varphi_{shift}}\left\{V_{in}\right\}\right) \tag{4}$$

3.2.3 *Optimisation of Bit Error Rate*

Following the modulation and amplification of the signal, the received signal is demodulated. The ΔDOR signal is extracted first. However, the concept of bit error rate does not apply to this signal. The TM and Rng signals are then extracted. As the

Rng signal is a known fixed sequence, the accepted bit error rate is greater than that for the TM data. In order to determine a bit error rate measure of the TM and Rng signals, we observed the individual spectral components. The measure of spectral interference of the TM signal on the Rng signal ($S_{Rng}I_{TM}R$) has been calculated for a range of modulation parameters. In addition, for the same modulation parameters the spectral interference of the Rng signal on the TM signal ($S_{TM}I_{Rng}R$) has been calculated. Both of these values are then minimised. The bandwidth of the transmitted signal is given as υ. Thus

$$S_{Rng}I_{TM}R(dB) = 10\log_{10}\left(\frac{\int_{\omega_c-\upsilon/2}^{\omega_c+\upsilon/2} \left| H_{Rng}(f) \right| df}{\int_{\omega_c-\upsilon/2}^{\omega_c+\upsilon/2} \left| H_{Rng}(f) \cap H_{TM}(f) \right| df} \right) \tag{5}$$

$$S_{TM}I_{Rng}R(dB) = 10\log_{10}\left(\frac{\int_{\omega_c-\upsilon/2}^{\omega_c+\upsilon/2} \left| H_{TM}(f) \right| df}{\int_{\omega_c-\upsilon/2}^{\omega_c+\upsilon/2} \left| H_{TM}(f) \cap H_{Rng}(f) \right| df} \right) \tag{6}$$

where

H_{TM} Power Spectrum of the TM signal
H_{Rng} Power Spectrum of the Rng signal

However, both $S_{Rng}I_{TM}R$ and $S_{TM}I_{Rng}R$ are functions of ω_{TM}, ω_{Rng}, Φ_{TM} and Φ_{Rng}. For optimising the transmission system, equation 7 holds, where *k* is a weighting factor.

$$S_{Rng}I_{TM}R(\omega_{TM}, \omega_{Rng}, \Phi_{TM}, \Phi_{Rng}) = kS_{TM}I_{Rng}R(\omega_{TM}, \omega_{Rng}, \Phi_{TM}, \Phi_{Rng}) \tag{7}$$

To simplify the determination of optimum spectral compatibility, the problem is broken down into a function of just two variables. The system is then optimised for two variables. These two variables are then fixed whilst a third is introduced, and so on until each of the modulation parameters have been included in the optimisation process. Figure 7 shows equation 7 solved for Φ_{TM}, Φ_{Rng} and k=1. The point labelled 'x' shows maximum spectral compatibility for the TM and Rng signals, with Φ_{TM}=0.8 and Φ_{Rng}=0.75.

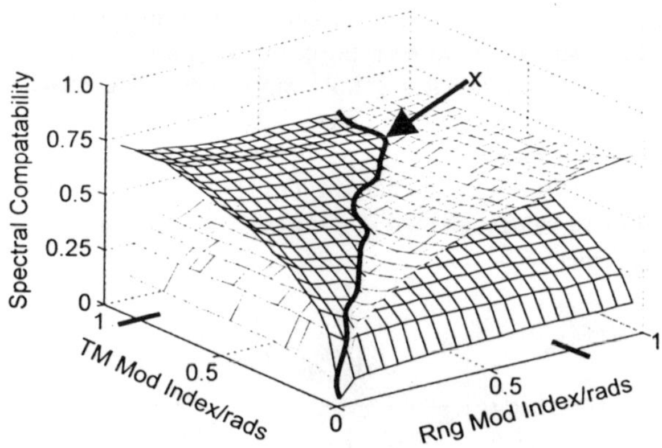

Figure 7 - Spectral Compatability

4. Conclusion

We have described techniques to simulate both a satellite communications system and a TWTA. We show how the techniques can be applied to a real deep space satellite to determine the set of optimum modulation parameters.

5. Acknowledgements

The work described in this paper was carried out under ESTEC contracts. The IMAL-2 simulations described in section 2. were carried out by Mr W. Bosch of ESTEC.

5. References

[1] P. G. Farrell, I. M. Arison, "Performance of an OFDM System in Frequency Selective Channels Using Reed-Solomon Coding Schemes", IEE Colloquim on Multipath Countermeasures, May 1996

[2] G. Gatti, W. Bosch, S. Colado Soriano, R. Martinez Ruiz de Gopegui and Y. Vichery, "IMAL-2, a simulation software for intermodulation analysis of power amplifiers", XRM Technical Notes, Volume V, Number 6, ESTEC, Noordwijk, The Netherlands, June 4, 1996.

[3] R.G. Carter, P. Wang, V. Srivastava and A.K. Sinha, "Computer-aides design tools for space TWTs", Proc. ESA Space TWTs 1994 Workshop, ESTEC, Noordwijk, The Netherlands, May 1994

[4] P. Young PE, "Electronic Communications Techniques", pp. 97, 1985

[5] S. Sonander, B. Honary, "Modulation Processes in Deep Space Transponder", Lancaster University, LCRC/01/08/3/SS/95, August 1995

[6] H. R. Schwart, "Numerical Analysis", John Wiley & Sons, ISBN 0-471-92065-7

[7] K. E. Atkinson, "An Introduction to Numerical Analysis", ARC A, ISBN 0-471-50023-2

18. Wireless ATM - A Review

A. H. Aghvami and B. Jafarian
Centre for Telecommunications Research
King's College London

Introductory Remarks

The future of modern telecommunications will be based on two distinct areas: the evolution of worldwide wireline infrastructures toward support for broadband multimedia services, and the revolution of cellular radio systems to support mobile users with a variety of services. Future fully developed Personal Communication System (PCS) technology must deliver an array of multimedia services to portable hand-held units at any desired location. Lightweight indoor/outdoor PCS terminals will allow tetherless access to the broadband network from home, office or even out of town [1].

Asynchronous Transfer Mode (ATM) as the next step in the evolution of switching systems, promises to create a multi-services network. It is the only solution for virtual networking in local and wide area networks. Proposed as the future infrastructure for B-ISDN by ITU-T, ATM has become the hottest topic in the telecommunications industry. Wireless ATM can be considered as a natural wireless extension to B-ISDN.

The Promise of Wireless ATM

The rapid penetration of cellular phones and portable PCs in both business and consumer markets is a proof that users place a significant value on portability. By introducing multimedia services into portable PCs and also migrating towards Personal Digital Assistance (PDAs), wireless extensions to broadband networks become more important. Such broadband wireless services could first start in the private Local Area Network (LAN) scenario, gradually moving to microcellular public PCS.

Wireless ATM, as an attractive candidate to provide such a system, is motivated by the growing acceptance of ATM as a standard for broadband networking applications. The concept of extending the standard ATM protocol over a wireless network interface was first proposed in [2,3], and has since been an active research topic in many organizations worldwide. All recent research activities indicate that it is possible to use standard ATM protocols to support seamless wired/wireless networking through the new wireless specific protocol sublayers (medium access control and logical link control) parallel to the ATM data plane. A limited mobility extension could also be added to the wired ATM network control plane. All of these would allow the extension of a multimedia network based on ATM into the wireless domain with end-to-end guaranteed QoS for fixed and mobile platforms.

In addition to the long-term scenario, as mentioned, there is a short-term scenario to use the current access systems (e.g. GSM, IS-54 and IS-95) to support mobility in ATM networks. The use of ATM networks as the backbone is particularly advantageous in microcell and integrated voice/data scenarios [4]. Table 1 shows the typical applications should be supported by wireless ATM networks [2].

Applications	Service Type	QoS	Bit-rate Range
Voice telephony	CO/CBR	Call blocking permitted; Medium cell loss; Isochronous	2.4 - 32 Kbps
Digital audio	CO/CBR	Call blocking permitted; Low cell loss; Low delay jitter	128 - 512 Kbps
Teleconference	CO/CBR or CO/VBR	Call blocking permitted; Medium cell loss; Low delay jitter	64 - 384 Kbps
Digital HDTV	CO/CBR	Call blocking permitted; Low cell loss; Low delay jitter	15 - 20 Mbps
General Computer Data	CL	No call blocking; Low cell loss; Medium delay and jitter	0.1 - 1 Mbps
E-mail	CL	No call blocking; Low cell loss; High delay	9.6 - 128 Kbps
High-speed data	CL	No call blocking; Very low cell loss; Medium delay and jitter	1 - 10 Mbps

Table 1 Typical application requirements for wireless ATM

Wireless ATM System Architecture

As mentioned before, wireless ATM is considered as the wireless extension of the B-ISDN infrastructure. This extension is in the sense of extending the user-access network to the wireless environment. The basic structure of the wireless ATM system is shown in Fig. 1. The system could support a wide range of services for users, with different terminal mobility in different environments.

The various system concepts for the wireless ATM system are as follow:

- Cellular Radio ATM networks: To extend limited B-ISDN services into a cellular mobile network, at least in highly crowded areas, with a capacity of 2 Mb/s at the air interface.

- Wireless ATM for Residential: Similar to the DECT system, especially for residential entertainment applications. The aim is to support operation of interactive television and multimedia PCs anywhere within a house without requiring a wired connection.

- Wireless ATM LANs: The extension of current standards (HIPERLAN I and IEEE 802.11) to support radio-based LAN-like communications for indoor environments with a capacity of 25-40 M/bs.

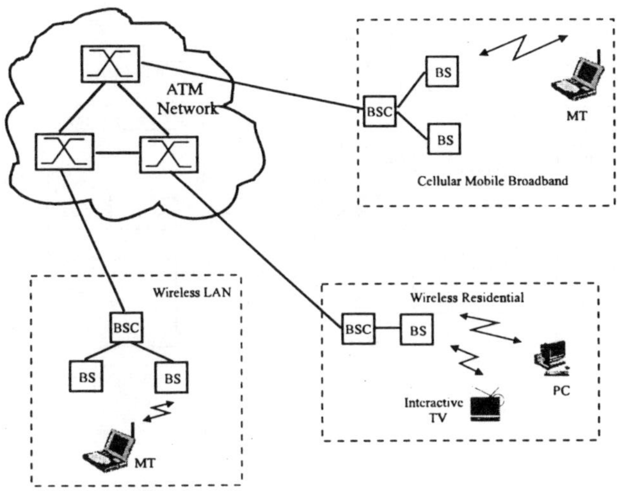

Fig. 1 Different access environment for Wireless ATM

The full integration of mobile ATM terminals into a fixed ATM network requires retransmission of ATM cells over the air interface in such a way that protocols of the ATM Adaptation Layer (AAL) are not involved. As shown in Fig. 2, the ATM physical layer is replaced by a radio-specific physical layer at the air interface. Like all other mobile cellular systems, each base station serving one radio cell, is connected to a base station controller. The functionality of the mobile physical layer (M-PHY) is mainly contained in each base and mobile station, whereas the protocol for the mobile medium access control (M-MAC) and mobile logical link control (M-LLC) sublayers are running peer-to-peer between base station controllers and mobile stations [5].

Transport Mechanism for Wireless ATM

Wireless ATM originates by applying the ATM-type transport mechanism through the radio link for future systems such as B-ISDN. The main advantage of ATM is flexible resource management for mixed traffic in the shape of different rate bearers and also guaranteed QoS for all of them. All of the bearers mentioned demand their own performance parameters, such as delay tolerance, delay jitter and bit-error rate. As proposed in [6], all the entities of the network layer protocol could map to five different classes inside the data link layer (AAL sub-layer) and then use resource sharing (M-MAC) to map the bearers onto the physical channels. Fig. 3 shows the general view of this mapping.

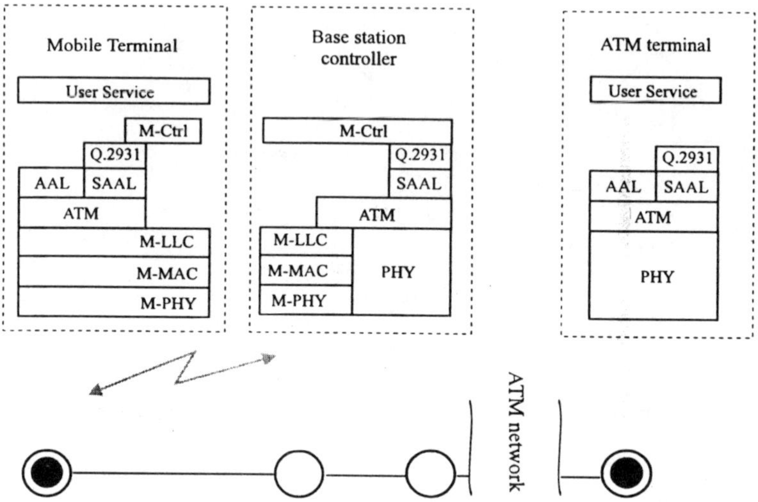

Fig. 2 Protocol stack for a wireless ATM system

The increase of wireless system capacity is the main issue of the transport mechanism in wireless ATM systems. Optimizing spectral efficiency for integrated narrow/broad-band traffic is the main task of the transport mechanism. The wireless channels are often divided into forward and reverse channels. Forward channels are used and controlled by the base station to transfer data to the mobile users. Reverse channels are shared by the mobile users for transmitting to the base station. In reverse channels a MAC protocol is often needed to coordinate channel access among users.

Radio Access Layer

It is shown in Fig. 2 that there is a radio access layer (including physical and a part of data link layers) which extends the ATM network into the wireless domain. This layer is divided into three sub-layers.

Mobile Radio Physical Layer (M-PHY): Wireless ATM requires a high-speed radio modem capable of providing reasonably reliable transmission in microcell and picocell environments. As mentioned before, the frequency bands will be decided by the environment, indoor or outdoor. The candidate modulation methods could be equalized QPSK/QAM [7].

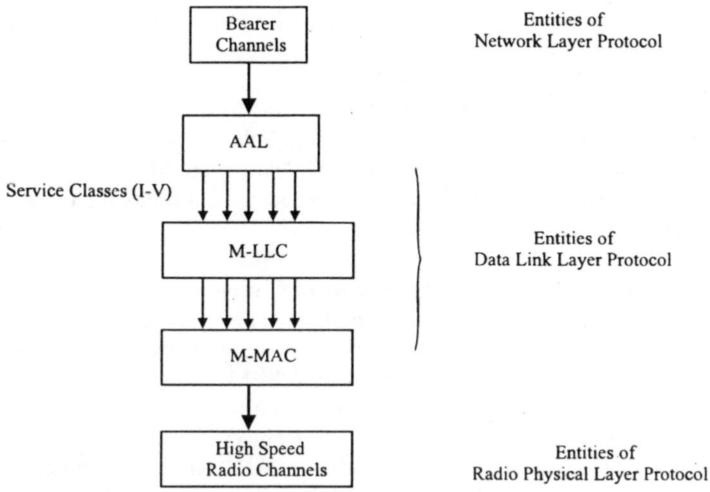

Fig. 3 Transport Entities and Layers for Wireless ATM

Mobile Medium Access Control (M-MAC) sub-layer: A MAC sub-layer is required to support shared use of radio channels by multiple terminals. The MAC sub-layer must provide support for standard multimedia services, such as Constant Bit Rate (CBR), Variable Bit Rate (VBR), Available Bit Rate (ABR) and Unspecified Bit Rate (UBR), traffic classes with associated QoS controls. A key factor in selection of the MAC protocol will be the ability to support different traffic classes at reasonable QoS levels while maintaining a reasonably high radio channel efficiency. Different protocols could be considered as the candidates for such a system, but the Random Reservation Access (RRA) family appears to be the main candidate for M-MAC.

- RRA with reservation compartment: A promising class of MAC protocols for wireless ATM access is that of RRA protocols with separate reservation compartment. In this family, each time slot is mapped onto a physical channel and time slots are grouped into equal size frames. In the uplink, one slot in each frame is dedicated for reservation [8]. Each terminal having a service to be transmitted tries to reserve a slot in the request slot. A reserved slot is exclusively used by the terminal that has reserved it. Examples of these protocols are ALOHA-Reservation and PRMA++. The advantages of this family are ability to perform dynamic channel allocation, high channel throughput and guaranteed QoS. The disadvantage of this family is waste of channel bandwidth (around 10%) because of the reservation slot.

- RRA without reservation compartment: R-ALOHA based protocols are members of this group. In these protocols, each terminal having a service to be transmitted contends for the next available time slot. Upon successful reception of the first packet the base station grants the terminal a reservation for exclusive use of the same time slot in each subsequent frame. PRMA is one of the protocols in this class.

Mobile Logical Link Control (M-LLC) sub-layer: The M-LLC protocol in conjunction with the network/transport layer serves to insulate higher layers from the effects of wireless channel impairments such as blocking and delay at the M-MAC sub-layer. The M-LLC protocol is a critical issue in wireless operation since it limits the propagation of errors into the core network.

The wireless medium will generally require a M-LLC protocol that interfaces with the network layer in a relatively transparent manner. This sub-layer is required to insulate the higher layer services from wireless medium specific problems caused by a high error rate at the physical layer and blocking and delay at the M-MAC level. The design of the M-LLC sub-layer takes into account the specific requirements of the services to be supported. The main requirement is the provision of powerful error detection/correction techniques and retransmission procedures. A variety of ARQ and FEC techniques exist in the literature, but support of real-time multimedia services in wireless ATM places additional requirements on the services provide by the M-LLC sub-layer.

Frame transmission in wireless ATM networks

As mentioned before, the main approach in wireless ATM protocol design is toward improving the reliability of transmission through wireless environment. This also includes the concept of cell encapsulation in the wireless environment. The main task of M-LLC, is to prepare the frames for transmission through the wireless environment with highest reliability and minimum overhead.

In the UNI, the 5 byte ATM header consists of 3 byte VCI and VPI for routing the cells to network. But we should realise that the routing capability in wireless ATM systems is only supported inside the wireless access protocol. The location management is within the MSC and is a part of the wireless access protocol. Because the logical connection between the mobile user and MSC is one-to-one correspondence, it is clear that the VCI and VPI information fields overlap with wireless routing information. As proposed in [2], the compression of the ATM header could reduce the overhead of the wireless frame. Fig. 4 shows the flow structure of frames in a wireless ATM network.

Mobility Management for Wireless ATM networks
There are three basic issues for handling the mobility of terminals in a wireless ATM network [9]:

- Location management
- Connection management
- Handoff management

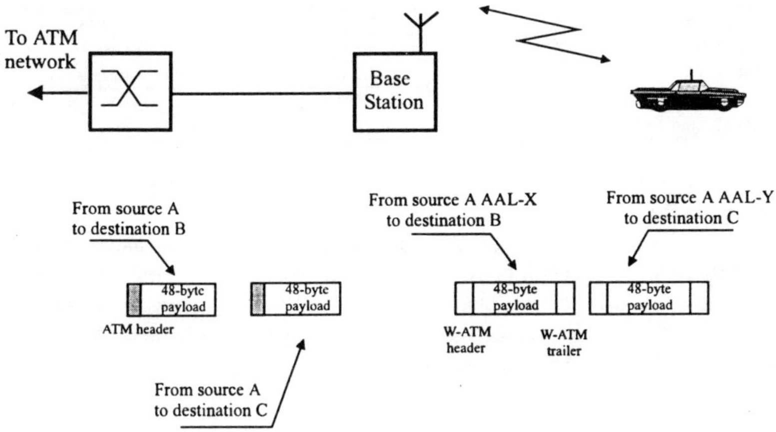

Fig. 4 Flow structure of frames in wireless ATM networks

Location management deals with tracking the position of a mobile and handling queries regarding the location of a mobile. For cellular extension the database concept of Home Location Register (HLR) and Visitor Location Register (VLR) , such as in GSM, could be useful. However, for LANs, other solutions based on broadcasting may be simpler to implement without requiring the overhead of a database concept. The BAHAMA project [3] uses the same concept for location management.

In connection-oriented mobile networks, such as wireless ATM, segments of the connection need to be torn down as the mobile user moves from one cell to another. Connection and handoff management issues deal with this problem. Maintaining the QoS during the handoff is the main issue in mobility management for wireless ATM networks. There are three rerouting categories for handoff [10]: anchor rerouting used in IS-41 and GSM cellular networks [11,12], dynamic rerouting [13], and virtual connection tree [14]. In the first two categories, segments of established connection are torn down and set up during handoffs. The last category avoids segment tear down and extra setups during handoffs by setting up extra segments of the connection tree (from a root switch to a set of base stations that mobile could roam during the connection) during the connection setup phase [15]. The relative advantages and disadvantages of the above three rerouting methods are summarized in Table 2 [10].

System parameters	Anchor rerouting	Dynamic rerouting	Virtual connection tree
Impact on wire-line signaling	None	Significant	Significant
Spare capacity and VPI/VCI	Least	Least	Most
Control complexity	Simplest	Most complex	Complex
Inter-MSC traffic	Most	Least	Least
Handoff delay	Minimum	Highest	Moderate
Can be supported by SS7 networks?	Yes	No	No

Concluding Remarks

ATM technology has become the only choice for future broadband integrated communications. A wired infrastructure for such a network will be implemented very soon. In this article we have tried to mention the main issues regarding the extension of such a system into the wireless mobile domain. It is clear that real and commercially available wireless ATM systems are not too far away.

References

[1] R. R. Gejji, "Mobile Multimedia Scenario Using ATM and Microcellular Technologies", *IEEE Trans. Veh. Technol., vol. 43, no 3, Aug. 1994*, pp. 699-703.

[2] D. Raychaudhuri and N Wilson, "ATM Based Transport Architecture for Multiservices Wireless Personal Commuications", *IEEE J. Select. Areas Commun., vol. 12, no. 8, Oct. 1994*, pp. 1401-1414.

[3] K. Y. Eng et. al., "BAHAMA: A Broadband Ad-Hoc Wireless ATM Local Area Network", *Proc. IEEE ICC'95*, pp. 1216-1223.

[4] D. Raychaudhuri, "Wireless ATM Networks: Architecture, System Design and prototyping", *IEEE Personal Commun. Mag., vol. 3, no. 4, Aug. 1996*, pp. 42-49.

[5] B. Walke, D. Petras and D. Plassmann, "Wireless ATM: Air Interface and Network Protocol of the Mobile Broadband System", *IEEE Personal Commun. Mag., vol. 3, no. 4, Aug. 1996*, pp. 50-56.

[6] F. Cheng and M. Holtzman, "Wireless Intelligent ATM Network and Protocol Design for Future Personal Communication Systems", *IEEE J. Select. Areas Commun., vol. 15, no. 7, Sep.. 1997*, pp. 1289-1307.

[7] R. Valenzuela, "Performance of Quadrature Amplitude Modulation for Indoor Radio Communications", *IEEE Trans. On Commun., vol. COM-35, no. 11, Nov. 1987*, pp. 1236-38.

[8] B. Jafarian and A. H. Aghvami, "Performance of Multiservice Dynamic Reservation (MDR) TDMA for Multimedia Wireless Networks", *Proc. IEEE ICUPC'96*, pp. 291-294.

[9] E. Ayanoglu, K. Y. Eng and M. J. Karol, " Wireless ATM: Limits, Challenges and Protocols", *IEEE Personal Commun. Mag., vol. 3, no. 4, Aug. 1996*, pp. 18-34.

[10] T. H. Wu and L. F. Chang. "Architectures for PCS Mobility Management on ATM Transport Networks", *Proc. IEEE ICUPC'95*, pp. 763-68.

[11] EIA/TIA IS-41 (Revision B), "Cellular Radio-Telecommunications Intersystem Operations", *Jan. 1992.*

[12] GSM Rec. 0.9.02, "Mobile Application Part (MAP) Specification".

[13] L. V. Hauwermelren et al., "Requirements for Mobility Support in ATM", *Proc. IEEE Globecom'94*, pp. 1691-1695.

[14] A. S. Acampora and M. Naghshineh, "An Architecture and Methodology for Mobile -Executed Handoff in Cellular ATM Networks", *IEEE J. Select. Areas Commun., vol. 12, no. 8, Oct. 1994*, pp. 1365-1375.

[15] C. Noblet, B. Jafarian and A. H. Aghvami, "Quality of Service and Overload Concepts in Wireless Multimedia Network Based on Cellular ATM", *Proc. IEEE ICC'97*, pp. 1458-1462.

19. Multifunctional Applications of Complementary Sequences in Radio Communication Systems

Michael Darnell
Institute of Integrated Information Systems
University of Leeds
Leeds LS2 9JT
UK

Abstract

The paper first discusses the theoretical basis of complementary sequence set synthesis, together with the important correlation properties of such sequence sets; binary, multilevel and complex sets are considered. The concept of multifunctionality is then reviewed and the way in which complementary sequence sets can be applied within a multifunctional radio communications system architecture is briefly described.

1. Introduction

The major practical importance of complementary sequences (CSs) in the communications system context lies in the fact that they offer useful correlation properties in both aperiodic and periodic modes. The former is especially valuable where burst-mode transmission is required.

Important initial contributions to the theory and synthesis of sets of CSs were made by (Welti, 1960), (Golay, 1961), (Turyn, 1963), (Schweitzer, 1971), (Tseng & Liu, 1972) and (Griffin, 1977); since that era, there has been a considerable body or research in the topic, for which a comprehensive bibliography can be found in (Fan & Darnell, 1996).

Section 2 presents the basic structures and synthesis procedures for binary CS sets, whilst Section 3 shows how these techniques can be extended to multi-level (>2) and complex sequence sets, having arbitrary dimensions. Section 4 then introduces the concept of multifunctionality, and discusses one example of the application of CSs within a multifunctional radio communications system design, whilst Section 5 presents overall conclusions.

2. Binary Complementary Sequence Sets: Structures and Synthesis

2.1 Binary Pairs of Complementary Sequences

Golay (Golay, 1961) defines a pair of binary, bipolar, CSs as "a pair of equal-length sequences in which the number of pairs of like elements with a given separation in one sequence is exactly equal to the number of pairs of unlike elements with the same separation in the other sequence". An example of such a pair is as follows:

$$\begin{array}{lrrrr} \text{Sequence 1:} & +1 & +1 & -1 & +1 \\ \text{Sequence 2:} & +1 & -1 & -1 & -1 \end{array} \qquad (1)$$

This definition leads directly to the characteristic aperiodic autocorrelation function (ACF) property of CSs, viz that the autocorrelation vector sum is zero everywhere, except at the in-phase (zero-shift) position. For the sequences of (1), the aperiodic ACF is computed as shown below

Shift (Units of $\Delta\tau$):	-3	-2	-1	0	1	2	3
Sequence 1 ACF:	+1	0	-1	+4	-1	0	+1
Sequence 2 ACF:	-1	0	+1	+4	+1	0	-1
Summed ACF:	0	0	0	+8	0	0	0

(2)

where $\Delta\tau$ is the clock interval for both sequences, which are assumed to be time-aligned.

In (Darnell, 1975), generalised synthesis procedures are given for binary CS pairs with sequences of length N bits, where

$$N = 2^n \text{ (n integer} \geq 1) \tag{3}$$

giving a zero-shift summed ACF peak of 2^{n+1}; these procedures will be outlined in Section 2.2.

It should be noted that the aperiodic crosscorrelation function (CCF) between the individual sequences in a given pair is non-zero; for example, the CCF for the sequences of (1) is of the form:

Shift:	-3	-2	-1	0	1	2	3
CCF:	-1	-2	-1	0	+1	-2	+1

(4)

Hence, it is important that, if the sequence pairs are to be applied in a manner that exploits the impulsive aperiodic ACF properties, they should be processed in a way which eliminates the effects of the non-zero CCF; the practical implications of this statement are discussed in Section 5.

In practice, the aperiodic CCF between distinct CS pairs of the same length is often important. Consider the two pairs of sequences shown in (5) below:

	Pair (i)		Pair (ii)
Sequences 1:	+1 +1 -1 +1	and	+1 +1 +1 -1
Sequences 2:	+1 -1 -1 -1	and	+1 -1 +1 +1

(5)

The aperiodic CCF between pairs (i) and (ii) is obtained by summing, at corresponding shifs, the separate aperiodic CCFs between sequences 1 and sequences 2, ie

Shift:	-3	-2	-1	0	1	2	3
Sequences 1 CCF:	+1	0	+1	0	+3	0	-1
Sequences 2 CCF:	-1	0	-1	0	-3	0	+1
Summed CCF:	0	0	0	0	0	0	0

(6)

It is seen that the summed aperiodic CCF is zero for all shifts; thus, the two pairs are said to be uncorrelated in a complementary sense.

It may be shown, eg in (Darnell, 1993), that aperiodic and periodic ACFs can be simply related by the following expression

$$\phi_p(s) = \phi_a(s) + \phi_a(s + N) \tag{7}$$

where $\phi_p(s)$ and $\phi_a(s)$ are respectively the periodic and aperiodic ACF values at discrete shift s, and N is the length of the sequence being correlated. Therefore, if $\phi_a(s) = 0$ for all $s \neq 0$, the corresponding $\phi_p(s)$ values will also be zero; this means that the "ideal" aperiodic ACF properties of CS pairs will be present when the same sequences are correlated periodically, eg for the binary CS pair

Sequence 1:	+1	+1	-1	+1	-1	-1	-1	+1	
Sequence 2:	+1	-1	-1	-1	-1	+1	-1	-1	(8)

the summed aperiodic and periodic ACFs are:

Shift s:	-7	-6	-5	-4	-3	-2	-1	0	1	2	3	4	5	6	7
Sequence 1 $\phi_a(s)$:	1	0	-3	0	-1	0	-1	8	-1	0	-1	0	-3	0	1
Sequence 2 $\phi_a(s)$:	-1	0	3	0	1	0	1	8	1	0	1	0	3	0	-1
Sum $\phi_a(s)$:	0	0	0	0	0	0	0	16	0	0	0	0	0	0	0
Sequence 1 $\phi_p(s)$:	0	0	-4	0	-4	0	0	8	0	0	-4	0	-4	0	0
Sequence 2 $\phi_p(s)$:	0	0	4	0	4	0	0	8	0	0	4	0	4	0	0
Sum $\phi_p(s)$:	0	0	0	0	0	0	0	16	0	0	0	0	0	0	0

(9)

2.2 Synthesis of Pairs of Binary Complementary Sequences

It is possible to extend the concept of complementary sequences to sets of more than two sequences. To show how this can be accomplished, it is first necessary to review synthesis techniques for CSs. In (Darnell, 1975), a method of synthesising CS pairs is described which uses sequence "block structures", as follows:

Sequence 1:	$+A\rightarrow$	$+B\rightarrow$	
Sequence 2:	$+B\leftarrow$	$-A\leftarrow$	(10)

where A and B are half-length (N/2) sequence blocks of length 2^{n-1} digits, the signs +/- indicate the relative polarity of all the digits in a half-length block, and the arrows $\rightarrow/\leftarrow$ indicate the sense in which all the digits in a half-length block should be read. As a simple synthesis example, take as arbitrary "seeds"

$$+A\rightarrow \;\equiv\; +1$$

$$+B\leftarrow \;\equiv\; -1 \qquad (11)$$

and, by employing the block structure of (10), recursions for n = 1, 2, 3,.... can be carried out:

Recursion for n = 1	+1	+1							
	+1	-1							(12)
Recursion for n = 2	+1	+1	-1	+1					
	+1	-1	-1	-1					(13)
Recursion for n = 3	+1	+1	-1	+1	-1	-1	-1	+1	
	+1	-1	-1	-1	-1	+1	-1	-1	(14)
etc.									

It is apparent that for n ≥ 2, the 1st halves of the sequences obtained from the ith recursion are themselves the complete sequences from the (i - 1)th recursion; the complete sequences obtained from the (i - 1)th recursion are defined as the blocks $+A\rightarrow$ and $+B\leftarrow$ for the ith recursion.

2.3 Synthesis of Sets of Binary Complementary Sequences

It is possible to extend the synthesis procedure of the previous section to CS sets of more than two sequences by redefining blocks A and B to represent CS sets, rather than individual sequences. For example, by taking the CS pair of (12), and redefining as

$$\begin{matrix} +1 & +1 \\ +1 & -1 \end{matrix} \equiv +A \text{ and } +B \tag{15}$$

then the block structure of (10) yields

Sequence 1:	+1	+1	+1	+1	
Sequence 2:	+1	-1	+1	-1	
Sequence 3:	+1	+1	-1	-1	(16)
Sequence 4:	-1	+1	+1	-1	

which is a (4 bit x 4 sequence) CS set. This can be verified by computing the aperiodic ACF properties:

Shift:	-3	-2	-1	0	1	2	3	
Sequence 1 ACF:	1	2	3	4	3	2	1	
Sequence 2 ACF:	-1	2	-3	4	-3	2	-1	(17)
Sequence 3 ACF:	-1	-2	1	4	1	-2	-1	
Sequence 4 ACF:	1	-2	-1	4	-1	-2	1	
Summed ACF:	0	0	0	16	0	0	0	

Since the summed aperiodic ACF vector is everywhere zero, apart from zero-shift, the above set may also be said to be complementary.

2.4 Uncorrelated Binary Complementary Sequence Sets

In some applications to be outlined in Section 5, it is essential to have available uncorrelated sets of CSs. One example of binary uncorrelated CS set synthesis will now be discussed, although alternative techniques are also available (Darnell & Kemp, 1992) (Fan & Darnell, 1996). Consider the two block structures shown below:

$$\begin{matrix} +A\rightarrow & +B\rightarrow \\ +B\rightarrow & -A\rightarrow \\ \multicolumn{2}{c}{\text{(i)}} \end{matrix} \qquad \begin{matrix} +A\rightarrow & -B\rightarrow \\ +B\rightarrow & +A\rightarrow \\ \multicolumn{2}{c}{\text{(ii)}} \end{matrix} \tag{18}$$

where the conventions are as used previously. Choosing

$$A = B = 1 \tag{19}$$

and applying the block structures if (18)(i) and (ii) gives two (2 x 2) binary sequence pairs

$$\begin{matrix} 1 & 1 \\ 1 & -1 \\ \multicolumn{2}{c}{\text{(i)}} \end{matrix} \quad \text{and} \quad \begin{matrix} 1 & -1 \\ 1 & 1 \\ \multicolumn{2}{c}{\text{(ii)}} \end{matrix} \tag{20}$$

It can be simply verified that the sets of (20)(i) and (ii) are both complementary and uncorrelated. If (20)(i) and (ii) are now redefined respectively as +A and +B, the block structures of (18)(i) and (ii) can be used recursively to produce two (4 x 4) uncorrelated CS sets, ie

$$\begin{matrix} 1 & 1 & 1 & -1 \\ 1 & -1 & 1 & 1 \\ 1 & -1 & -1 & -1 \\ 1 & 1 & -1 & 1 \\ & \text{(i)} & & \end{matrix} \quad \text{and} \quad \begin{matrix} 1 & 1 & -1 & 1 \\ 1 & -1 & -1 & -1 \\ 1 & -1 & 1 & 1 \\ 1 & 1 & 1 & -1 \\ & \text{(ii)} & & \end{matrix} \tag{21}$$

If the two uncorrelated sets of (20) are now juxtaposed to form a single array **D**, where

$$\mathbf{D} = \begin{matrix} 1 & 1 & 1 & -1 \\ 1 & -1 & 1 & 1 \end{matrix} \tag{22}$$

then an extended array **D'** can be defined by (Tseng & Liu, 1972)

$$\mathbf{D'} = \begin{matrix} \mathbf{D/D} & \mathbf{-D/D} \\ \mathbf{-D/D} & \mathbf{D/D} \end{matrix} \tag{23}$$

where **-D** indicates that all the elements of **D** are negated, and the symbol / denotes alternate interleaving of columns of the two arrays specified. Hence, **D'** becomes

$$\begin{matrix} 1 & 1 & 1 & 1 \\ 1 & 1 & -1 & -1 \\ -1 & 1 & -1 & 1 \\ -1 & 1 & 1 & -1 \\ & \text{(i)} & & \end{matrix} \quad \begin{matrix} 1 & 1 & -1 & -1 \\ 1 & 1 & 1 & 1 \\ -1 & 1 & 1 & -1 \\ -1 & 1 & -1 & 1 \\ & \text{(ii)} & & \end{matrix} \quad \begin{matrix} -1 & 1 & -1 & 1 \\ -1 & 1 & 1 & -1 \\ 1 & 1 & 1 & 1 \\ 1 & 1 & -1 & -1 \\ & \text{(iii)} & & \end{matrix} \quad \begin{matrix} -1 & 1 & 1 & -1 \\ -1 & 1 & -1 & 1 \\ 1 & 1 & -1 & -1 \\ 1 & 1 & 1 & 1 \\ & \text{(iv)} & & \end{matrix} \tag{24}$$

Each of the (4 x 4) sets (i) - (iv) in (24) are complementary and uncorrelated in a complementary sense with the other three sets. By redefining (24) as **D**, expression (23) can be used recursively to generate a (64 x 8) array, which can then be partitioned into say eight (8 x 8) uncorrelated CS sets; alternatively, only a subset of (24), eg (i) and (ii), can be defined as **D** to produce a (32 x 8) array of four (8 x 8) uncorrelated sets. Clearly, other partitionings are also possible.

2.5 Binary Complementary Sequence Set Operations

Certain operations can be carried out on a binary CS set, without affecting its complementary properties; these are:

(a) interchanging any number of sequences in the set;
(b) reversing any number of sequences in the set;
(c) negating any number of sequences in the set;
(d) negating alternate elements of all sequences in the set.

For example, given the set (24)(i) above:

$$\begin{matrix} 1 & 1 & 1 & 1 \\ 1 & 1 & -1 & -1 \\ -1 & 1 & -1 & 1 \\ -1 & 1 & 1 & -1 \end{matrix} \tag{25}$$

then the following (4 x 4) CS sets can be obtained by carrying out the four operations (a) - (d) above:

$$
\underset{\text{(a)}}{\begin{matrix} 1 & 1 & 1 & 1 \\ -1 & 1 & -1 & 1 \\ -1 & 1 & 1 & -1 \\ 1 & 1 & -1 & -1 \end{matrix}}
\qquad
\underset{\text{(b)}}{\begin{matrix} 1 & 1 & 1 & 1 \\ -1 & -1 & 1 & 1 \\ 1 & -1 & 1 & -1 \\ -1 & 1 & 1 & -1 \end{matrix}}
\qquad
\underset{\text{(c)}}{\begin{matrix} -1 & -1 & -1 & -1 \\ 1 & 1 & -1 & -1 \\ 1 & -1 & 1 & -1 \\ -1 & 1 & 1 & -1 \end{matrix}}
\qquad
\underset{\text{(d)}}{\begin{matrix} 1 & -1 & 1 & -1 \\ 1 & -1 & -1 & 1 \\ -1 & -1 & -1 & -1 \\ -1 & -1 & 1 & 1 \end{matrix}}
\qquad (26)
$$

2.6 Synthesis of Odd-Length Binary Complementary Sequence Sets

Thus far, only CS sets with lengths given by 2^n (n integer $\geq$ 1) have been synthesised. It is not possible to produce binary CS sets with an odd number of sequences per set; this can be seen immediately by considering the initial 1-digit overlap condition in the aperiodic ACF computation, where an odd number of sequences will always give a non-zero ACF value. However, CS sets having an even number of sequences, but odd-length sequences, can readily be synthesised.

As an example, consider the (8 x 4) CS set below:

$$
\begin{matrix} 1 & 1 & -1 & 1 & 1 & 1 & -1 & 1 \\ 1 & -1 & -1 & -1 & 1 & -1 & -1 & -1 \\ 1 & -1 & 1 & 1 & -1 & 1 & -1 & -1 \\ -1 & -1 & -1 & 1 & 1 & 1 & 1 & -1 \end{matrix}
\qquad (27)
$$

Shorter sequence length sets can be derived by selecting subsets of (27), eg

$$
\underset{\text{(i)}}{\begin{matrix} 1 & 1 \\ 1 & -1 \\ 1 & -1 \\ -1 & -1 \end{matrix}}
\qquad
\underset{\text{(ii)}}{\begin{matrix} 1 & 1 & -1 \\ 1 & -1 & -1 \\ 1 & -1 & 1 \\ -1 & -1 & -1 \end{matrix}}
\qquad
\underset{\text{(iii)}}{\begin{matrix} 1 & -1 & 1 \\ -1 & -1 & -1 \\ 1 & -1 & -1 \\ 1 & 1 & -1 \end{matrix}}
\qquad
\underset{\text{(iv)}}{\begin{matrix} 1 & -1 & 1 & 1 & 1 \\ -1 & -1 & -1 & 1 & -1 \\ -1 & 1 & 1 & -1 & 1 \\ -1 & -1 & 1 & 1 & 1 \end{matrix}}
\qquad
\underset{\text{(v)}}{\begin{matrix} -1 & 1 & 1 & 1 & -1 & 1 \\ -1 & -1 & 1 & -1 & -1 & -1 \\ 1 & 1 & -1 & 1 & -1 & -1 \\ -1 & 1 & 1 & 1 & 1 & -1 \end{matrix}}
\qquad (28)
$$

All the sets (28)(i) - (v) above are complementary, but (ii), (iii) and (iv) are of odd length.

In general, similar selection procedures can be employed with starting arrays of different dimensions.

3. Multilevel and Complex Complementary Sequence Sets

Many of the synthesis procedures which have been described previously in the context of binary CS sets can be directly applied to non-binary, multi-level (>2) CS set synthesis.

3.1 Multilevel Sets of Complementary Sequences

The block structures of Section 2.2 can be applied to non-binary, real sequences. Taking A = 1 and B = 2, and applying to (18)(i) gives:

$$
\begin{matrix} 1 & 2 \\ 2 & -1 \end{matrix}
\qquad (29)
$$

which is a (2 x 2) CS pair. If (29) is now redefined as being equal to A and B, using (18)(i) again yields the (4 x 4) sequence array:

$$
\begin{matrix} 1 & 2 & 1 & 2 \\ 2 & -1 & 2 & -1 \\ 1 & 2 & -1 & -2 \\ 2 & -1 & -2 & 1 \end{matrix}
\qquad (30)
$$

which is a 4-level CS set. Now, taking the block structure of (18)(ii) with A = 3 and B = 4, we obtain

$$\begin{array}{rr} 3 & -4 \\ 4 & 3 \end{array} \qquad (31)$$

If (29) is defined as A and (31) as B, then (18)(i) gives the 7-level CS set:

$$\begin{array}{rrrr} 1 & 2 & 3 & -4 \\ 2 & -1 & 4 & 3 \\ 3 & -4 & -1 & -2 \\ 4 & 3 & -2 & 1 \end{array} \qquad (32)$$

Uncorrelated multilevel real CS sets can be generated in the same way as binary sets; similarly, odd-length sets can be derived by selection of subsets from larger arrays. The same techniques can also be employed to generate real CS sets with non-integer elements.

One important extension to the CS library, not achievable with binary sets, becomes feasible using multilevel elements, ie CS sets with an odd number of sequences; the block structures below can be used for this purpose (Darnell & Kemp, 1992):

$$\begin{array}{rr} A & -2A \\ -2A & A \\ -A & -4A \\ \\ \\ \multicolumn{2}{c}{(i)} \end{array} \qquad \begin{array}{rrr} 2A & A & -2A \\ A & 2A & 2A \\ 2A & -2A & A \\ \\ \\ \multicolumn{3}{c}{(ii)} \end{array} \qquad \begin{array}{rr} A & -2A \\ -2A & A \\ -A & -3A \\ -4A & -A \\ -A & 3A \\ \multicolumn{2}{c}{(iii)} \end{array} \qquad \begin{array}{rrr} A & 2A & -2A \\ -2A & -A & A \\ -A & -3A & -3A \\ -4A & A & -A \\ -A & -3A & 3A \\ \multicolumn{3}{c}{(iv)} \end{array} \qquad (33)$$

Again, the above structures can be expanded recursively by using (18) or (33); an example of this makes use of (33)(i) twice, initially with A = 1, ie

$$\begin{array}{rrrr} 1 & -2 & -2 & 4 \\ -2 & 1 & 4 & -2 \\ -1 & -4 & 2 & 8 \\ -2 & 4 & 1 & -2 \\ 4 & -2 & -2 & 1 \\ 2 & 8 & -1 & -4 \\ -1 & 2 & -4 & 8 \\ 2 & -1 & 8 & -4 \\ 1 & 4 & 4 & 16 \end{array} \qquad (34)$$

which can be shown to be a (4 x 9) CS set.

3.2 Complex Sets of Complementary Sequences

It is also possible to synthesise complex (polyphase) CS sets, again using some of the same techniques as described previously for real-valued sequences. Examples of pairs of 4-phase sequences with elements s_i are (Fan & Darnell, 1996):

$$\begin{array}{ccccc} 0 & 1 & 3 & 2 & 1 \\ 0 & 0 & 0 & 1 & 3 \end{array} \qquad (35)$$

and

$$\begin{array}{cccccccccc} 0 & 3 & 0 & 3 & 2 & 1 & 0 & 1 & 2 & 1 \\ 0 & 3 & 0 & 1 & 2 & 3 & 0 & 3 & 0 & 3 \end{array} \qquad (36)$$

where the corresponding CS pair elements $\{s_{ih}\}$ are obtained from the mapping

$$\{s_i\} = \exp(j2\pi s_i)/4 \qquad (37)$$

A (5 x 4) quadriphase CS set can be obtained, using the mapping of (37), from

$$\begin{matrix} 0 & 1 & 2 & 1 & 0 \\ 1 & 0 & 3 & 0 & 3 \\ 3 & 0 & 3 & 0 & 1 \\ 0 & 1 & 2 & 1 & 0 \end{matrix} \qquad (38)$$

Similarly, the elements s_i of a (3 x 3) set of 3-phase sequences are given by

$$\begin{matrix} 0 & 0 & 0 \\ 0 & 1 & 2 \\ 0 & 2 & 1 \end{matrix} \qquad (39)$$

In this case, the corresponding CS elements $\{s_i\}$ are obtained from the mapping

$$\{s_i\} = \exp(j2\pi s_i)/3 \qquad (40)$$

There is still further research to be carried out on the more general synthesis of both individual and uncorrelated sets of CSs with specified dimensions and numbers of phases. Also, comparatively little consideration has been given to complex CS sets where the elements do not take constant magnitude values, with regularly-spaced phases.

4. Multifunctionality

4.1 The Nature of Multifunctional Coding

Multifunctional coding (Darnell & Honary, 1993) is an attempt to coalesce in a systematic manner the various elements of channel evaluation, encoding and decoding to achieve both a more powerful algorithmic structure and a more efficient DSP-based implementational architecture. In the context of this paper, only multifunctional coding applicable to radio systems is considered.

The original multifunctional (MF) concept (Darnell & Honary, 1986) encompassed the following functions:

(a) error control encoding/decoding;
(b) multi-user encoding/decoding;
(c) synchronisation;
(d) modulation/demodulation;
(e) security;
(f) real-time channel evaluation (RTCE).

The RTCE, or channel state, data is required to initiate appropriate adaptation, both algorithmic and architectural, in response to significant changes in the parameters of the radio path being used. To allow such adaptation to take place rapidly, it is assumed that a software radio configuration is employed.

4.2 A Multifunctional Transmission Procedure Employing Complementary Sequence Sets

It is possible to use uncorrelated sets of CSs as a flexible vehicle for adaptive MF data transmission over radio channels. In recent years, there has been increasing interest in the multi-channel Orthogonal Frequency-Division Multiplex (OFDM) scheme for radio channels because of its inherent resistance to multipath and selective fading effects. It should be noted, however, that the OFDM technique is not new and was first employed for HF (2 - 30 MHz) long-range ionospheric radio communications in the early 1950s. It will be shown that CS sets form a natural partner for the OFDM format, and can form the basis of an extremely robust transmission scheme (Prasad & Darnell, 1991).

Consider the multiple channel OFDM format shown below:

	Sub-channel 1 (f_1):	N digits
Frequency ↑	Sub-channel 2 (f_2):	N digits
	Sub-channel 3 (f_3):	N digits
	.	
	.	
	Sub-channel M (f_M):	N digits
		→ Time

It is evident that an (N x M) CS set could be mapped directly on to the above configuration; the first sequence in the set would modulate the carrier of sub-channel 1 at frequency f_1, the second sequence would modulate the carrier of sub-channel 2 at frequency f_2, etc. It can be shown that of the number of sequences in a CS set is M, then a maximum of (M - 1) additional CS sets, uncorrelated in a complementary sense with the first, can be synthesised (Fan & Darnell, 1996). Thus, data can be transmitted by mapping data words to CS sets and their inverses, with detection being accomplished by appropriate matched filtering. For example, if M = 16, then 5bits of information per CS set can be transmitted.

The CS set dimensions can be chosen to suit the prevailing channel state. The energy of each information bit transmitted is effectively dispersed in time and frequency over the complete (N x M) CS set array, hence providing simultaneous time and frequency diversity to combat fading and narrowband noise/interference, together with controllable transmission redundancy. The capacity of the system can be increased by overlaying uncorrelated CS sets, possibly offset in time, and adding corresponding elements in the array to produce composite elements with more discrete states.

In summary, the multiple simultaneous (MF) capability of the OFDM - CS combination provides the following features:

(a) adaptive error control due to the inherent redundancy of the transmission scheme, and the ability to select CS sets of different dimensions;

(b) controllable time and frequency diversity via adaptation of the dimensions of the CS array;

(c) simultaneous multiple-user access, since different users can be assigned distinct uncorrelated CS sets for their information transmission (essentially, the scheme provides either aperiodic or periodic code-division multiple-access (CDMA));

(d) self-synchronisation, due to the impulsive nature of the CS set ACFs which can be exploited in a matched filter-based word synchroniser (note all sets used will have the same "perfect" ACF);

(e) a capability for continously evaluating the impulse response function of the channel by input-output crosscorrelation, using a test signal with an impulsive ACF.

Clearly, the optimum adaptation of the CS set parameters for any given channel state poses a significant analytical and practical problem.

5. Concluding Remarks

This paper has briefly reviewed synthesis procedures for, and properties of, a range of binary, real multilevel and complex CS sets. Methods of generating perfectly uncorrelated sets of CSs have also been indicated.

An important application area, ie multifunctional radio system design employing CS sets, has also been described.

There remains considerable scope for further research into the synthesis of CS sets with specified properties, particularly complex sets which can be mapped directly to complex modulation schemes.

References

[1] G.R. Welti. Quaternary codes for pulsed radar. *IRE Trans. on Information Theory*, IT-6: 400 - 408, June 1960.

[2] M.J.E. Golay. Complementary series. *IRE Trans. on Information Theory*, IT-7: 82 - 87, April 1961.

[3] R. Turyn. Ambiguity functions of complementary sequences. *IEEE Trans. on Information Theory*, IT-9: 46 - 47, January 1963.

[4] B.P Schweitzer. *Generalised Complementary Code Sets*. PhD thesis, University of California, Los Angeles, USA, 1971.

[5] C.C. Tseng and C.L. Liu. Complementary sets of sequences. *IEEE Trans. on Information Theory*, IT-18(5): 644 - 652, September 1972.

[6] M. Griffin. There are no Golay complementary sequences of length 2.9^t. *Aequationes Mathematicae*, 15: 73 - 77, 1977.

[7] P. Fan and M. Darnell. *Sequence Design for Communications Applications*, chapter on Complemetary sequences: 311 - 336. Communications Systems, Techniques and Applications Series, Research Studies Press ISBN 0 86380 201 X, 1996.

[8] M. Darnell. Principles and applications of binary complementary sequences. *Symposium on Theory and Applications of Walsh Functions*, Hatfield, 1975.

[9] M. Darnell. *Perturbation Signals for System Identification*, chapter entitled Periodic and non-periodic, binary and multi-level pseudorandom signals: 176 - 208. Prentice-Hall (K.R. Godfrey Ed.), ISBN 0 13 656414 3, 1993.

[10] M. Darnell and A.H. Kemp. *Cryptography and Coding II*, chapter entitled Multilevel complementary sequence sets: synthesis and applications: 37 - 67. The IMA Conference Series. Clarendon Press, Oxford (C. Mitchell, Ed.), ISBN 0 19 853393 4, 1992.

[11] M. Darnell and B. Honary. *Cryptography and Coding III*, chapter entitled Multi-functionality and its application in digital communication system design: 129 - 149. The IMA Conference Series. Clarendon Press, Oxford (M.J. Ganley, Ed.), ISBN 0 19 853691 7, 1993.

[12] M. Darnell and B. Honary. Multi-functional coding schemes applicable to secure communication systems. *2nd IEE International Conference on Secure Communications*. IEE CP-269: 74 - 82. London, 1986.

[13] K.V. Prasad and M. Darnell. Data transmission using complementary sequence sets. *5th IEE International Conference on HF Radio Systems and Techniques*. IEE CP-339: 222 - 226. Edinburgh, 1991.

20. A Code in the Nose

(In fond memory of Edward C. Posner)

Rodney M. Goodman

Department of Electrical Engineering,
Computation and Neural Systems,
California Institute of Technology (136-93),
Pasadena, CA 91125, U.S.A.

email: rogo@caltech.edu
WWW: http://www.micro.caltech.edu/micro/Goodman.html
Phone: 818 395 6740, Fax: 818 795 4765.

Abstract

At the time of his death Ed Posner was working on a paper with others at Caltech - the working title of which was "A Code in the Nose". Ed's lead has developed into a vigorous multi-disciplinary effort at Caltech to understand olfaction and to build the "Silicon Nose." I thought that all of you (and Ed indirectly) might enjoy hearing about our latest theories on olfactory coding, hypotheses, hardware, and not the least where olfaction is concerned, puns.

Developing a functional model of the mammalian olfactory sense is arguably the greatest intellectual frontier in chemical sensing, and electronic information processing. Electronic processing in the areas of vision, audition, touch, locomotion, and indeed specific chemical sensing are all well established. General olfaction remains elusive for several reasons: in particular the lack of realistic biological models, the availability of practical chemical sensors that function in a manner similar to biological receptors, and the development of neuromorphic silicon architectures that model neural processing. Our research at Caltech addresses each of these issues in the context of both biological and silicon olfaction. This multi-disciplinary approach spans neurobiology, chemistry, and electronic engineering. We are conducting research in biological insight and modeling of olfactory processing and coding, into neuromorphic VLSI hardware to implement signal processing and learning, and have developed for the *first* time broadly-tuned polymer chemical sensor arrays that emulate mammalian olfactory receptor neurons. We are modeling olfaction via realistic software neuronal models, animal experimentation, *and* in neuromorphic VLSI hardware systems. This results in a unique symbiotic relationship between biology and engineering, where biological insight and modeling can drive engineering silicon design, and where a practical electronic system can in turn pose new science questions to biology, which may result in revised directions in modeling, theory, or animal experimentation. The results of this will further drive the engineering design, and so on. Our ultimate goal is an understanding of the biological architecture of general olfaction, and just as importantly the construction of a single chip neuromorphic electronic silicon nose capable of odor classification and location.

Our silicon nose utilizes an array of polymer sensors to transduce the odor signal into a multi-spectral electrical signal, which then utilizes adaptive neuromorphic signal processing to condition those signals, and neural network pattern recognizers to classify the smell. In addition we are building robots that use the nose to track an odor to its source. Our approach to building the silicon nose is that of Neuromorphic Engineering. The vision of Neuromorphic Engineering is a future where machines sense, interact with, learn from, and adapt to their environment with the same ease as living creatures do now. This vision can not be realized entirely with the current digital paradigm which can not provide the necessary capability in a cost-effective, low-power, miniaturized implementation. The basic tenet of the neuromorphic approach is that a necessary ingredient to realizing this grand vision is to incorporate architectures and methods inspired by biological sensory and nervous systems. This approach is manifested in the word "neuromorphic." We follow the lead of nature by developing special purpose systems for specific sensory processing tasks. By trading off generality, we achieve high performance sensory systems that are low cost, require low power, and can be miniaturized. To realize this vision, we utilize powerful multi-disciplinary technologies that include low power analog VLSI implementations, integrated micromechanical sensors and actuators, artificial neural network learning algorithms and architectures, event driven communication, and integration of learning and adaptation into algorithms and hardware. This new discipline, pioneered by Carver Mead and other researchers at Caltech, and originally manifested in the Silicon Retina project, has roots that can be found in the age-old desire to learn from biology in which structure and function are intimately related. To summarize there are essentially three axioms of neuromorphic engineering that we follow in our methodology: observe and learn from biology; integrate adaptation and learning at all levels; and integrate the design of algorithms and hardware.

In my talk I'll discuss what we know about biological olfaction, and describe our silicon nose architecture and chips. There will be videos, demos, and things to smell! As the old music hall joke goes: "I've just invented an artificial nose!". "Really - How does it smell?" "Terrible!".